Introduction
to
Random
Differential
Equations
and
their
Applications

Modern Analytic
and Computational
Methods *in* Science
and Mathematics

A GROUP OF MONOGRAPHS
AND ADVANCED TEXTBOOKS

Richard Bellman, EDITOR
University of Southern California

Published

1. R. E. Bellman, R. E. Kalaba, and Marcia C. Prestrud, Invariant Imbedding and Radiative Transfer in Slabs of Finite Thickness, 1963

2. R. E. Bellman, Harriet H. Kagiwada, R. E. Kalaba, and Marcia C. Prestrud, Invariant Imbedding and Time-Dependent Transport Processes, 1964

3. R. E. Bellman and R. E. Kalaba, Quasilinearization and Nonlinear Boundary-Value Problems, 1965

4. R. E. Bellman, R. E. Kalaba, and Jo Ann Lockett, Numerical Inversion of the Laplace Transform: Applications to Biology, Economics, Engineering, and Physics, 1966

5. S. G. Mikhlin and K. L. Smolitskiy, Approximate Methods for Solution of Differential and Integral Equations, 1967

6. R. N. Adams and E. D. Denman, Wave Propagation and Turbulent Media, 1966

7. R. L. Stratonovich, Conditional Markov Processes and Their Application to the Theory of Optimal Control, 1968

8. A. G. Ivakhnenko and V. G. Lapa, Cybernetics and Forecasting Techniques, 1967

9. G. A. Chebotarev, Analytical and Numerical Methods of Celestial Mechanics, 1967

10. S. F. Feshchenko, N. I. Shkil', and L. D. Nikolenko, Asymptotic Methods in the Theory of Linear Differential Equations, 1967

11. A. G. Butkovskiy, Distributed Control Systems, 1969

12. R. E. Larson, State Increment Dynamic Programming, 1968

13. J. Kowalik and M. R. Osborne, Methods for Unconstrained Optimization Problems, 1968

14. S. J. Yakowitz, Mathematics of Adaptive Control Processes, 1969

15. S. K. Srinivasan, Stochastic Theory and Cascade Processes, 1969

16. D. U. von Rosenberg, Methods for the Numerical Solution of Partial Differential Equations, 1969

17. R. B. Banerji, Theory of Problem Solving: An Approach to Artificial Intelligence, 1969

18. R. Lattès and J.-L. Lions, The Method of Quasi-Reversibility: Applications to Partial Differential Equations. Translated from the French edition and edited by Richard Bellman, 1969

19. D. G. B. Edelen, Nonlocal Variations and Local Invariance of Fields, 1969

20. J. R. Radbill and G. A. McCue, Quasilinearization and Nonlinear Problems in Fluid and Orbital Mechanics, 1970

21. W. Squire, Integration for Engineers and Scientists, 1970

22. T. Parthasarathy and T. E. S. Raghavan, Some Topics in Two-person Games, 1971

23. T. Hacker, Flight Stability and Control, 1970

24. D. H. Jacobson and D. Q. Mayne, Differential Dynamic Programming, 1970

25. H. Mine and S. Osaki, Markovian Decision Processes, 1970

26. W. Sierpiński, 250 Problems in Elementary Number Theory, 1970

27. E. D. Denman, Coupled Modes in Plasmas, Elastic Media, and Parametric Amplifiers, 1970

28. F. H. Northover, Applied Diffraction Theory, 1971

29. G. A. Phillipson, Identification of Distributed Systems, 1971.

30. D. H. Moore, Heaviside Operational Calculus: An Elementary Foundation, 1971

32. V. F. Demyanov and A. M. Rubinov, Approximate Methods in Optimization Problems, 1970

33. S. K. Srinivasan and R. Vasudevan, Introduction to Random Differential Equations and their Applications, 1971

34. C. J. Mode, Multitype Branching Processes: Theory and Applications, 1971

37. W. T. Tutte, Introduction to the Theory of Matroids, 1971

In Preparation

31. S. M. Roberts and J. S. Shipman, Two-Point Boundary Value Problems: Shooting Methods

35. R. Tomović and M. Vukobratović, General Sensitivity Theory

36. J. G. Krzyż, Problems in Complex Variable Theory

38. B. W. Rust and W. R. Burrus, Mathematical Programming and the Numerical Solution of Linear Equations

39. N. Buras, Scientific Allocation of Water Resources: Water Resources Development and Utilization—A Rational Approach

40. H. M. Lieberstein, Mathematical Physiology: Blood flow and Electrically Active Cells

Introduction to Random Differential Equations and their Applications

S. K. Srinivasan
Indian Institute of Technology
Madras

R. Vasudevan
Institute of Mathematical Sciences
Madras

American Elsevier
Publishing Company, Inc.

NEW YORK · 1971

AMERICAN ELSEVIER PUBLISHING COMPANY, INC.
52 Vanderbilt Avenue, New York, N.Y. 10017

ELSEVIER PUBLISHING COMPANY, LTD
Barking, Essex, England

ELSEVIER PUBLISHING COMPANY
335 Jan Van Galenstraat, P.O. Box 211
Amsterdam, The Netherlands

International Standard Book Number 0–444–00097–6

Library of Congress Card Number 70–152756

Copyright © 1971 by American Elsevier Publishing Company, Inc.

Manufactured in Scotland

CONTENTS

CHAPTER 6
Stochastic Problems in Continuum Mechanics

CHAPTER 7
Transport Phenomena

CHAPTER 8
Path Integrals in Classical and Quantum Physics

PREFACE

At the beginning of this century, Albert Einstein and Smoluchowsky, in the course of the study of the motion of suspended pollen particles, introduced random equations and showed how the results of random walk theory could be used to describe the phenomenon. It was about the same time that Langevin proposed a random equation describing the motion of Brownian particles. A year or two later, Campbell, dealing with the fluctuations in the emission of rays by a radioactive substance, predicted the shot effect and proved a couple of theorems relating to the mean square behavior of the cumulative response due to such emissions. These instances mark the birth of stochastic integration and differential equations. Within a very short period, Wiener, anticipating Kolmogorov's formalization of probability, undertook the mathematical analysis of Brownian motion. Since then, random integrals and differential equations have continued to attract the attention of mathematicians, physicists, and electrical engineers. Although many specific types of random equations have been studied in the past fifty years, it is only recently that concerted attempts have been made to arrive at a formal, yet useful, theory of random equations. In a comprehensive survey presented at a symposium of the Royal Statistical Society in 1949, Moyal gave a detailed account of the theory of random integrals and equations with examples lucidly drawn from different disciplines of the physical sciences. This was followed by a physical approach to stochastic processes and, in particular, to stochastic integrals by Ramakrishnan, who gives an excellent account of the subject in his survey article published in *Handbuch der Physik*. About the same time, Blanc-Lapierre and Fortet in their encyclopedic treatment of random functions highlighted many of the physical phenomena that can be interpreted as cumulative response to random pulses. Since then, there has been a tremendous amount of activity, notably among physicists and electrical and structural engineers, apart from the systematic study of stochastic integrals by mathematicians. The objective of this monograph is to give a systematic account of some of these developments with special reference to their applications in different disciplines of physics and engineering.

Chapter 1 introduces the necessary preliminary concepts of stochastic processes. The educated probabilist will find it superfluous and perhaps annoying. We suggest that he need glance through it only to familiarize himself with the level (!) of rigor and the notation employed in the remainder of the book. However, we would like to caution the aspirant probabilist that this chapter cannot replace a preparatory course in stochastic processes. In Chapter 2 we introduce integrals of a class of random functions. We then discuss stochastic differential equations of different types. Chapter 4 contains an account of the theory of Brownian motion responsible for the deep impact on the research work of physicists and mathematicians. Chapter 5 deals with response phenomena, while Chapter 6 is devoted to the stochastic differential equations encountered in the realms of continuum mechanics. Since problems of transport phenomena are becoming increasingly important, we found it worthwhile to devote Chapter 7 to propagation in stochastic media. The final chapter contains an account of the path integral formalism, the importance of which we need hardly stress.

In the choice of the material we have been primarily motivated by those processes that have impressed us by their impact on the development of the theory and their relevance to applications. We have not hesitated to include a full discussion of any isolated problem we have come across whenever it held promise of potential from the viewpoint of either general theory or further applications. Nor do we claim exhaustiveness in the coverage of the subject matter or in the References. Questions like stochastic convergence of integrals, stability, and optimization have been deliberately avoided since an excellent account of them can be found in the work of Stratonovich published earlier in this series.

The reader is assumed to have a knowledge at the level of Feller's *Probability Theory*, Volume I. A knowledge of Ramakrishnan's *Probability and Stochastic Processes* appearing in Volume 3 of *Handbuch der Physik* may be advantageous. Throughout the book the viewpoint is that of an Applied Probabilist, and we have freely resorted to phenomenological methods whenever we have found it safe and advantageous. Since we have laid emphasis on the mode of applications rather than on abstract theory, we expect the monograph to be used by graduate students specializing in Applied Probability, Statistical Physics, and Continuum Mechanics. Specialists in Vibration and Structure Theory may also find the book useful.

We are deeply indebted to our former teacher, Professor Alladi Ramakrishnan, who introduced this subject to us, and to Professor Richard Bellman who encouraged us to launch this project.

We appreciate the help of Dr. N. V. Koteswara Rao in the preparation of the manuscript. We are also thankful to S. M. Rajasekharan for typing the manuscript and preparing it for the printer and to Miss. S. Kalpakam for help in correcting the proofs and preparing the index.

S. K. SRINIVASAN
R. VASUDEVAN

Madras
January, 1970

Chapter 1

RANDOM PHENOMENA

1. INTRODUCTION

... the main charm of probability theory lies in the enormous variety of its applications. Few mathematical disciplines have contributed to as wide a spectrum of subjects, a spectrum ranging from number theory to physics, and even fewer have penetrated so decisively the whole of our scientific thinking.

MARK KAC

These remarks of Kac [1] amply project the central role of modern probability theory in the unraveling of the mysteries of nature whether for purposes of explanation or prediction. In addition to its deep impact in physical sciences, the theory of random phenomena plays a vital role in other realms of thought ranging from social and political sciences to biological situations; the evolution of the human being or of an economic situation, the events controlling the growth of a colony of cells, the development and evolution of our species on this earth, and the intricate mechanism of the collective behavior of the innumerable number of neurons of the brain, besides a vast array of technological problems, constitute only a small sample of the variety and scope of the study of random processes. A natural mathematical tool to describe such fluctuating phenomena is the random function defined on a suitable parametric space. The prediction of the future of the system, though possible only in a certain statistical sense, is achieved by the study of the evolution of the random function or functions with reference to the characteristic parameter. If the properties of these random functions are completely unpredictable in the true sense of the term, no consistent theory at all can be formulated. However, in many of these phenomena a certain regularity even in their randomness can be discerned, a regularity which can be identified with certain statistical properties. This important feature enables us to extend to the problems of random phenomena the methods of analysis usually employed for the study of fully determinate systems. Thus the random functions obey difference or differential or integro-differential equations characteristic of their evolution. Although such equations, called random equations, have been studied in the past by many workers, only recently have attempts been made to develop the theory of random equations. In this book we present an account of some of the attempts and techniques of handling such random equations. This chapter, introductory in nature, deals with some of the fundamental notions of stochastic processes of which repeated use is made in the subsequent chapters.

In Section 2 we introduce stochastic processes and summarize some of the main results pertaining to both continuous and point processes. The next section contains a brief account of the distribution functions and the differential equations satisfied by them. We devote the final section to a general discussion of phenomena describable by random equations.

2. STOCHASTIC PROCESSES

A stochastic process, according to Doob [2], is simply a probability process; that is, any process in nature whose evolution we can analyze successfully in terms of probability. Perhaps the basic difference between the older mathematical probability and that of the years since about 1930 lies in the importance laid on the concept of a chance variable or a random variable. The random variable can be defined as a positive-valued function defined on the sample space. The subtle distinction between classical probability theory and stochastic process arises from the notion of a random function which is nothing but a random variable corresponding to the sample space whose elements correspond to functions (real or complex) (see Blanc-Lapierre and Fortet [3]). There is an alternative way, due to Doob [2] (see also Loeve [4]), of defining a stochastic process as a family of random variables x_t (or $x(t)$) indexed by $t \in T$. The family is an ordered set if t is one-dimensional. The variation of t over the index set T introduces a dynamical element into our problem, and hence the behavior of the random variable for various values of t can be visualized as some (physical) process with some random element in its structure. No difficulty is experienced in defining a process even when, at a particular t, we have an aggregate $x_1(t), x_2(t), \ldots$ since t can be varied in each of these. However, the interesting feature of $x(t)$ as distinguished from any aggregate lies in the fact that it is possible to represent many physical processes involving some random element by establishing correspondence between $x(t)$ or a suitable aggregate $x_1(t), x_2(t), \ldots$ and such processes.

For our purposes we can follow Ramakrishnan [5] and divide stochastic processes broadly into four classes, taking the case where we have one random variable at any particular t:

(i) both x and t discrete,

(ii) x discrete and t continuous,

(iii) x continuous and t discrete,

(iv) both x and t continuous.

Corresponding to these cases, we can define the probability frequency functions by

(i) $\pi(x_i; t_j)$, where $\pi(x_i; t_j)$ denotes the probability that $x(t_j) = x_i$, assuming that $x_1, x_2, \ldots, t_1, t_2, \ldots$ are permissible values of x and t, respectively,

(ii) $\pi(x_i; t)$, where $\pi(x_i; t)$ denotes the probability $x(t) = x_i$,

(iii) $\pi(x; t_j)$, where $\pi(x; t_j) \, dx$ denotes the probability that $x(t_j)$ takes a value between x and $x + dx$,

(iv) $\pi(x; t)$, where $\pi(x; t) \, dx$ denotes the probability that $x(t)$ takes a value between x and $x + dx$.

The processes described above can be divided into two classes according as x is or is not capable of assuming positive integral values only. Processes falling under the former class are known as point processes, a terminology due to Wold [6, 7], for, in such a case, $x(t)$ can be identified with events or incidences represented as points along the t-axis. Such processes have also been recognized by Kendall [8], Bhabha [9], and Ramakrishnan [10] in their study of distribution of discrete number of entities distributed over a continuous parameter. In fact, the processes studied by these authors turned out to be more general than the class anticipated by Wold in that the processes are evolutionary. It is easy to see that the processes falling in classes (i) and (ii) above can be made to satisfy the definition of the point process by a relabeling of the range of discrete values that the random variable assumes. We can make a further classification among (i) and (ii) and call them discrete point processes (for example, see Srinivasan [11,12]) and continuous point processes, respectively. On the other hand, we shall call the processes falling under categories (iii) and (iv) continuous processes.* For both categories of processes there exists a special case in which the probability structure is independent of the choice of the origin of t. Such processes are called stationary processes. To make our ideas clear, let us consider category (iv). In this case the stochastic process is completely specified by the sequence of joint probability frequency functions $\pi_m(x_1, t_1; x_2, t_2; \ldots; x_m, t_m)$, $m = 1, 2, \ldots$, where $\pi_m(x_1, t_1; x_2, t_2; \ldots; x_m, t_m) \, dx_1 \, dx_2 \ldots dx_m$ denotes the joint probability that $x(t_k)$ lies between x_k and $x_k + dx_k$, $k = 1, 2, \ldots, m$. Thus stationarity implies that π_m is invariant for arbitrary t translations for each m. Very often the parameter t signifies time, and in such a case the origin is

---

* Wold [6] prefers to call them ordinary processes.

usually chosen at $-\infty$. If, however, t does not signify time and stands for spatial point, say in three-dimensional Euclidean space, the process is better known as homogeneous rather than stationary. We shall encounter such processes in Chapter 4 when we deal with response of continuous structures subject to random forces which are homogeneous in space and stationary in time. Sometimes we may not visualize the evolution of the process during the infinite past; in such a case, stationarity would mean invariance under arbitrary but restricted finite time translation.

There is another criterion through which a dichotomy can be introduced. For any parametric value t and arbitrarily small $h > 0$, the probability structure corresponding to the parametric value $t + h$ may depend only on the probability structure corresponding to the parametric value t. In such a case, $x(t)$ is called a Markov (or Markovian) process; non-Markovian processes consist of the residuary class whose members do not satisfy the requirement above. We shall see presently how the Markov property can be used to obtain the probability frequency function (hereinafter referred to as p.f.f.) governing $x(t)$.

2.1. Continuous Processes

In this subsection we summarize some of the concepts and results that are in current use in the study of stationary continuous random functions. Since we introduce time averages, we assume ergodicity to go from time averages to expectations. Let us consider the random process $x(t)$ for a very long time T and assume that $x(t)$ is zero outside the time interval T. Then the function $x(t)$ is developed as a Fourier integral given by

$$x(t) = \int_{-\infty}^{+\infty} df A(f) e^{2\pi i f t}, \tag{2.1}$$

where $A(f) = A^*(f)$. Using Parseval's theorem and the fact that $|A(f)|^2$ is an even function of f, it follows by the ergodic hypothesis that

$$\mathbf{E}\{x^2\} = \lim_{T\to\infty} \frac{1}{T} \int_{-\infty}^{+\infty} x^2(t)\, dt = \int_0^\infty df G(f), \tag{2.2}$$

where

$$G(f) = \lim_{T\to\infty} \frac{2}{T} |A(f)|^2 \tag{2.3}$$

is called the spectral density. More generally, we have the result due to Wiener [13]:

$$\mathbf{E}\{x(t)x(t + \tau)\} = \int_0^\infty df G(f) \cos 2\pi f\tau. \tag{2.4}$$

Thus in any problem we need to find the spectral density corresponding to $x(t)$. This result will be useful especially when we have to deal with integrals or iterated integrals of $x(t)$.

A very special stochastic process, known as the Gaussian random process, has been studied in great detail in the past and has proved to be useful in the description of physical phenomena (see, for example, Wang and Uhlenbeck [14]). To introduce this process we assume that $x(t)$ is repeated periodically with the period T, where T is so very large that $x(t)$ can be developed as a Fourier series:

$$x(t) = \sum_{k=1}^{\infty} (a_k \cos 2\pi f_k t + b_k \sin 2\pi f_k t) \qquad (2.5)$$

where $f_k = k/T$. We have assumed that the average value of $x(t)$ is zero without loss of generality. The Gaussian random process is defined to be the one in which the coefficients a_k, b_k of the Fourier series defined by (2.5) are statistically independent of each other and are Gaussianly distributed with

$$\sigma_k{}^2 = \overline{a_k{}^2} = \overline{b_k{}^2} = \frac{G(f_k)}{T}. \qquad (2.6)$$

There is another equivalent manner of characterizing a Gaussian random process by imposing the following restrictions on the correlations of $x(t)$:

$$E\{x(t_1)x(t_2) \ldots x(t_{2n+1})\} = 0,$$

$$E\{x(t_1)x(t_2) \ldots x(t_{2n})\} = \sum_{\text{all pairs}} E\{x(t_i)x(t_j)\}E\{x(t_k)x(t_l)\} \ldots, \qquad (2.7)$$

where the sum is to be taken over all the different ways in which we can divide the $2n$ points into n pairs.

2.2. Point Processes

As mentioned in the introductory remarks of this section, a point process is defined by the random variable $x(t)$, where $x(t)$ is concerned with the events or incidences which may be represented as points on the time axis. Although, in the original formulation of Wold, the parameter t is restricted to play the role of time (or is one-dimensional in nature), the efforts of Kendall [8], Bhabha [9], and Ramakrishnan [10] clearly indicate the general (multidimensional) nature of the parameter t. In fact, the formulation has been used by Thompson [15], who dealt with such processes in multidimensional spaces in connection with some ecological problems. The most appropriate way of characterizing the process is by the introduction of the random variable $N(t)$ denoting the number of events or incidences that have occurred up to t. A natural way to deal with the random variable $N(t)$ is by the study

2

of the characteristic functional introduced by Le Cam [see 11]. $C(\theta)$ is given by

$$C(\theta) = \mathbf{E}\{\exp i \int_0^\infty \theta(t) \, dN(t)\}. \tag{2.8}$$

If we are in possession of $C(\theta)$, we can determine all the statistical properties of the point process. However, explicit determination of $C(\theta)$ is extremely difficult except in one or two special instances (for example, see [11, Chap. 2, Sec. 4]). In view of this, Wold [7] and a few others like Cox and Smith [16] and McFadden [17] confined themselves to the study of the distribution of consecutive events occurring on the t-axis by introducing $h_1(t_0, t_1)$, $h_2(t_0, t_1, t_2), \ldots, h_n(t_0, t_1, t_2, \ldots, t_n), \ldots$, where* $h_n(t_0, t_1, t_2, \ldots, t_n) \, dt_1 \, dt_2 \ldots dt_n$ denotes the probability that, given an event has occurred at t_0, the ith event (excluding the one at t_0) occurs between t_i and $t_i + dt_i$ $i = 1, 2, \ldots, n$. To make further progress, we can impose stationarity. Thus we can choose t_0 as origin and replace $h_n(t_0, t_1, t_2, \ldots, t_n)$ by $h_n(u_1, u_2, \ldots, u_n)$, where $u_i = t_i - t_0$. Next we observe that $h_n(u_1, u_2, \ldots, u_n)$ obeys the condition

$$\int_0^\infty du_1 \int_{u_1}^\infty du_2 \ldots \int_{u_{n-1}}^\infty du_n h_n(u_1, u_2, \ldots, u_n) = 1. \tag{2.9}$$

A very special class of processes where the intervals between successive events are independently and identically distributed are known as renewal processes. In such a case, $h_n(u_1, u_2, \ldots, u_n)$ is given by

$$h_n(u_1, u_2, \ldots, u_n) = h_1(u_1)h_1(u_2 - u_1) \ldots h_1(u_n - u_{n-1}), \tag{2.10}$$

a result which brings to light that $h_1(u)$ completely determines the characteristics of the process. A renewal process can be conceived of as a non-Markovian point process whose memory extends only up to the point of occurrence of the previous event (or incidence). Of course, if $h_1(u)$ is an exponential function of u, then the point process is Markovian and can be identified with a homogeneous Poisson process. Wold [6] also defined a generalized renewal process in which the memory extends up to the kth previous event, and such processes are expected to be of great significance in the analysis of neuronal firings. However, progress is seriously limited if we proceed along this line of thought. The reason is that, in a general non-Markovian situation, we need the functions h_n of all orders to arrive at experimentally observable statistical properties of the system. We may have to seek functions corresponding to renewal densities (for example, see Cox

---

* The utility of the functions h_n seems to have been noticed by Janossy [18] in an entirely different context and, perhaps, was well known to kinetic theorists as early as 1937 (see, for example, de Boer [19]).

[20]) in nonrenewal situations. A year after Wold's publication of his results, such functions were introduced in a natural way by Ramakrishnan [10] while dealing with continuous parametric systems. As we have observed earlier, the formalism of the latter has a decided advantage over that of Wold because Wold imposed stationarity right from the start. A critical account of the generalized renewal processes in such a formalism and some of the recent developments can be found in the survey article by Srinivasan [21]. We now give a brief account of some of the results due to Ramakrishnan [10].

2.3. Product Densities and Correlation Functions

The central quantity of interest in Ramakrishnan's formalism is the random variable $dN(u)$, where $N(u)$ is the number of entities with parametric values less than or equal to u, u being a continuous parameter. If $p(m)$ = probability $\{dN(u) = m\}$, it is assumed that

$$p(1) = f_1(u)\, du + o(du),$$
$$p(m) = o(du) \qquad (m > 1),$$
$$p(0) = 1 - f_1(u)\, du + o(du). \tag{2.11}$$

Equations (2.11) imply that every one of the moments of the random variable $dN(u)$ is equal to $p(1)$. With the help of this assumption, it is easy to obtain the moments of $N(v)$. For instance, the mean value of $N(v)$ is given by

$$\mathbf{E}\{N(v)\} = \mathbf{E}\left\{\int_0^v dN(u)\right\} = \int_0^v f_1(u)\, du, \tag{2.12}$$

a relation which brings home the idea that, while $f_1(u)$ has a probability density interpretation, the integral of $f_1(u)$ over u yields only the mean number distributed over the interval of integration. An equivalent method of interpreting such a situation, due to Bartlett [22], consists in observing that only probability densities can be attached to particular values of u and not nonzero probabilities. Bartlett considers this a fitting reason to call such processes point processes.

It is obvious that we shall encounter expectation of products of $dN(u)$ if we proceed to calculate the higher moments of $N(v)$, which are defined as product densities* by Ramakrishnan [10]. The product density of degree 2 is defined by

$$\mathbf{E}\{dN(u_1)\, dN(u_2)\} = f_2(u_1, u_2)\, du_1\, du_2, \tag{2.13}$$

where du_1 and du_2 do not overlap. The function $f_1(u)$ is defined as the product density of degree 1. Next we observe that $f_2(u_1, u_2)\, du_1\, du_2$ denotes the

---

* *Note added in proof:* Correlation functions identical with the product densities were first introduced by S. O. Rice (*Bell. Syst. Tech. Journ.* **23** (1944) 282) in connection with the zero crossings of random processes. Although Rice did not use any special name for these functions, he studied them and obtained $h_1(t)$ as the sum of weighted integrals of product densities of different orders.

probability that an event occurs between u_1 and $u_1 + du_1$ and one occurs between u_2 and $u_2 + du_2$ irrespective of other events in the parametric range under consideration. This interesting property, which is shared by product densities of higher order, is the main reason for the wide applicability of product densities.

Kendall [8] independently defined similar correlation functions. Instead of the expectation value of the products of $dN(u)$, he dealt with the covariance of such products and defined them as cumulant densities. A detailed account of the product densities and their extensions as well as their connection with cumulant functions is given in the survey article by Srinivasan [21].

The problems considered by Wold and others can be elegantly handled by the introduction of product densities. In fact, we can characterize the point process by the family of product densities $f_m(u_1, u_2, \ldots, u_m)$, $m = 1, 2, \ldots$. This is quite reasonable because the product densities are nothing but the functional derivatives of the characteristic functional defined by (2.8). If we assume that the process is stationary, then

$$
\begin{aligned}
f_1(u) &= \text{a constant} = \alpha, \\
f_2(u_1, u_2) &= \alpha g_1(u_2 - u_1), \\
f_n(u_1, u_2, \ldots, u_n) &= \alpha g_{n-1}(u_2 - u_1, u_3 - u_1, \ldots, u_n - u_1).
\end{aligned}
\tag{2.14}
$$

Renewal processes can be defined as those satisfying the condition

$$
g_n(u_1, u_2, \ldots, u_n) = g_1(u_1)g_1(u_2 - u_1) \ldots g_1(u_n - u_{n-1}), \qquad n = 2, 3, \ldots,
\tag{2.15}
$$

where α is to be identified as the limit of $g_1(u)$ as u tends to infinity. Returning to the general case, we notice that the Fourier transform of $g_1(u)$ with respect to u multiplied by α will yield the power spectrum. This is the decided advantage of product densities. Finally, we note that the product density of degree 1 is nothing but the generalization of the renewal density for general point processes.

3. DISTRIBUTION FUNCTIONS AND DIFFERENTIAL EQUATIONS

So far, we have introduced different types of distribution functions for continuous as well as point processes and have discussed their general properties. However, the discussion has been confined only to the "geometrical" or "kinematical" aspects of the processes. In order to obtain structural relations, it is necessary to introduce some kind of "dynamical" assumption characterizing the particular process. For example, the Markov property introduced in Section 2 is a typical "dynamical" property. In this section we propose to demonstrate the possibility of arriving at structural relations which, in turn, determine the probability structure of the system either in terms of the p.f.f. or the product densities or other suitable correlation functions.

3.1. Markov Processes: Chapman-Kolmogorov Equations

Let us consider a Markovian process of category (iv) introduced in Section 2 and let $\pi(x|x_0, t)$ be the p.f.f. of $x(t)$ conditional on $x(t)$ having assumed the value x_0 at $t = 0$. The function $\pi(x|x_0, t)$ is completely determined by its behavior in the neighborhood of the parametric origin. For instance, if we assume that

$$\pi(x|x_0, h) = R(x|x_0)h + o(h) \tag{3.1}$$

for arbitrarily small $h > 0$, we can proceed to determine $\pi(x|x_0, t)$ for arbitrary t. To demonstrate this, let us assume further that the process is homogeneous with respect to t. Equation (3.1) implies that the random variable (if it assumed the value x_0 at $t = 0$) continues to take the same value x_0 with probability

$$1 - h \int_x R(x|x_0)\, dx + o(h).$$

Thus, using the homogeneous nature of the process, we can relate $\pi(x|x_0, t + h)$ to $\pi(x|x_0, t)$ by

$$\pi(x|x_0, t + h) = \left\{ 1 - h \int_{x'} R(x'|x)\, dx' \right\} \pi(x|x_0, t) +$$

$$h \int_{x'} \pi(x'|x_0, t)R(x|x')\, dx' + o(h). \tag{3.2}$$

Proceeding to the limit as h tends to zero, we obtain the Chapman-Kolmogorov equation (see Feller [23])

$$\frac{\partial \pi(x|x_0, t)}{\partial t} = -\pi(x|x_0, t)\int_{x'} R(x'|x)\, dx' + \int_{x'} \pi(x'|x_0, t)R(x|x')\, dx' \tag{3.3}$$

with the initial condition

$$\pi(x|x_0, 0) = \delta(x - x_0). \tag{3.4}$$

Equations (3.3) and (3.4), in principle, determine $\pi(x|x_0, t)$ for arbitrary t.

If the process is not homogenous with respect to t, the transition probability density $R(x|x')$ will depend on t as well. With this modification, (3.3) is still valid. If we find a process to be non-Markovian, in many instances we can define a new process involving some additional elements which render the process Markovian. In fact, non-Markovian processes can be visualized to arise from the projection of a Markovian process, the non-Markovian nature arising from the absence of extra "coordinates" lost by projection. We shall have occasion to illustrate this point in Chapters 2 and 3 when we deal with random integrals.

3.2. Imbedding Equations

The imbedding technique is a mathematical tool specially designed for the study of both Markovian and non-Markovian stochastic processes. The technique is an extension of certain invariance principles used in the study of reflection and transmission of light by piles of plates. The notion of imbedding has been used by Kolmogorov (see Feller [23]), particularly in his derivation of backward differential equations. Following the notation used in Section 3.1, we note that $\pi(x|x_0, t + h)$ can be expressed in terms of $\pi(x|x_0; t)$ by analyzing the outcome of possibilities in the interval $(0, h)$. Thus we have, for arbitrarily small $h > 0$,

$$\pi(x|x_0, t + h) = \left\{ 1 - h \int_{x'} R(x'|x_0) \, dx' \right\} \pi(x|x_0, t) +$$

$$h \int_{x'} R(x'|x_0)\pi(x|x', t) \, dx' + o(h). \qquad (3.5)$$

Equation (3.5) can be interpreted in an entirely different manner. By letting t have arbitrary values, we have imbedded the process corresponding to the interval $(0, t)$ into a class of processes. This point was first observed by Bellman and Harris [24], who extended this technique to a non-Markovian situation by dealing with age-dependent branching phenomena. They demonstrated the possibility of dealing with a non-Markovian process of a certain "duration" by imbedding it into a class of processes of arbitrary duration, the result of such an imbedding yielding a functional integral equation. This was developed further by Bellman, Kalaba, and Wing [25–27] in their comprehensive study of neutron transport theory.

The importance of the imbedding technique lies in its capability to yield integral or differential equations in the case of product densities defined over t space, for in such a case, in virtue of the functions being densities in t space, the traditional method of studying the changes in the interval $(t, t + h)$ fails completely. The success of this technique was extensively demonstrated by Srinivasan [11, 21]. We shall have occasion to use this technique in Chapter 4 when we deal with Brownian motion.

4. INTEGRATED RESPONSE TO STOCHASTIC PHENOMENA

It is a common experience that many physical situations which exhibit reasonable statistical features viewed on a large time scale are caused by the occurrence of a large number of highly random events. The temporal fluctuations of these random events may be quite large during short ranges of time, though their accumulated effects over macroscopic times have stable predictable equilibrium properties. The extreme molecular chaos in an ordinary gas existing in the microscopic domain, leading to macroscopic features of stable average values of density, pressure, and other macroscopic features, is an illustrative example. In shot noise and Barkhausen noise problems, the microscopic random events like the arrival of an electron on the anode or the sudden turnings of internal magnetic fields, respectively, occur at random time points t_i and give rise to specified effects governed by a response function at a later time T. This accumulated response is the sum of a number of stochastic contributions and hence, in dealing with these processes, we are naturally led to the concept of stochastic integration. By the same token, the evolution of such systems may be described by differential or integral or integro-differential equations which may be called random equations since the coefficients or the source term in these equations may be governed by a probabilistic structure. The mean and moments of the solutions of such equations, which represent the long time response of the system in response to stochastic causation, are of great interest in many fields of human activity. In dealing with such phenomena, we have to develop notions of stochastic calculus. Without going into details, we refer here briefly to some ideas of stochastic convergence and continuity for the sake of completeness.

Extending the usual ideas of convergence to random functions, we have to think of three major types of convergence reflecting the probabilistic structure of these functions. Taking a series of random variables $x_1, x_2, \ldots, x_n$, analogous to the three modes of convergence in ordinary analysis we have the following definitions:

(i) Almost certain convergence (a.c. convergence). When n tends to infinity, if every realized value x_n^R converges to x^R, x_n is said to converge in a.c. to the random variable x with probability 1.

(ii) Convergence in the mean square (m.s. convergence). x_n is said to converge to x in m.s. if, for arbitrary $\varepsilon > 0$,

$$\mathbf{E}\{|x_n - x|^2\} < \varepsilon, \qquad n \geqslant N. \tag{4.1}$$

(iii) Convergence in probability. If, for every $\varepsilon > 0$, we have

$$\text{Prob.}\,[|x_n - x| > \varepsilon] < \varepsilon \quad \text{for} \quad n \geqslant N, \tag{4.2}$$

x_n is defined to converge to x in probability.

These definitions provide successively weaker conditions for convergence and impose a structure on the covariance and correlations of the random variable (see, for example, Moyal [28], Ramakrishnan and Vasudevan [29]).

From the idea of convergence, the concept of continuity is immediate. The random function $x(t)$ is continuous in mean square at t if

$$\lim_{h \to 0} \mathbf{E}\{[x(t+h) - x(t)]^2\} = 0. \tag{4.3}$$

This is the starting point for carrying over to the domain of stochastic theory all the notions of differentiability and integrability, and also of various types of measures associated with integration, familiar in ordinary analysis (see Moyal [28]). In a series of papers, Ramakrishnan [30] elaborated on the integrals of random functions from a phenomenological viewpoint, stressing the fact that integrals of random functions are easier to handle than derivatives, and he used a pragmatic approach to the study of these equations, thus avoiding measure theoretic formulations.

In describing physical processes, we write differential equations wherein the coefficients are quantities measured as an average of a set of experiments. This may be an adequate description in many cases. However, if the fluctuations of these coefficients are large, it will be more realistic to introduce these quantities as random variables and obtain the "random solutions" and consider their statistical features. There are many models, such as Langevin equations, which require a separate stochastic source term in the differential equations. In addition, the coefficients and the source terms may be correlated. In general, a random linear differential equation may be of the form

$$Ly(t) = a_n(t)\frac{d^n y}{dt^n} + a_{n-1}(t)\frac{d^{n-1} y}{dt^{n-1}} + \ldots + a_0(t)y = x(t), \tag{4.4}$$

where the functions $a_i(t)$ and $x(t)$ may be stochastic in nature. The study of such equations from the viewpoint of functional analysis was initiated by the Prague school of probabilists. However, recently a host of new concepts and methods, such as perturbation techniques, stochastic Green's functions, diagrammatic analysis, concepts of trajectories and path probabilities, and integration over function spaces and stochastic operator formalisms, have been devised which are relevant and useful in particular situations. These concepts are described in various chapters in this book.

Random equations occur naturally in the description of a wide range of phenomena besides the well-known one of Brownian motion. The following are only some of the contexts in which random equations play a significant part: wave propagation in stochastic media, magnetohydrodynamics when the fields fluctuate, electrical circuit theory, turbulence theory, statistical mechanics of continuous medium, molecular scattering of light, certain magnetic resonance phenomena, astrophysical situations, transport problems, kinetic equations in physical chemistry, analysis of neuronal networks, problems in genetics, population studies, and bacterial growth. It is natural therefore that a book cannot cover the entire field and bring into the fore all the aspects of the innumerable number of individual physical problems. Hence we have limited ourselves to a few of the problems mentioned above which have interested us because of either the novelty of their equations or their physical significance.

REFERENCES

1. M. Kac, *Lectures in Probability and Related Topics in Physical Sciences*, Interscience, New York, 1959.
2. J. L. Doob, *Amer. Math. Monthly*, **49**(1942), 648.
3. A. Blanc-Lapierre and R. Fortet, *Théorie des Fonctions Aléatoires*, Masson et Cie., Paris, 1953. Translated by J. Gani, Gordon and Breach, New York, 1965, p. 81.
4. M. M. Loève, in *Lectures on Modern Mathematics*, Vol. III, T. L. Saaty, ed., John Wiley, New York, 1965, p. 245.
5. Alladi Ramakrishnan, Probability and Stochastic Processes, in *Handbuch der Physik*, Vol. 3, *Band III*, Springer-Verlag, Berlin, 1959.
6. H. Wold, *Skand. Aktuarietiotskr.*, **31**(1948), 229.
7. H. Wold, in *Le Calcul des Probabilités et Applications*, C.N.R.S., Paris, 1949, p. 75.
8. D. G. Kendall, *J. Roy. Statist. Soc.*, **B11**(1949), 230.
9. H. J. Bhabha, *Proc. Roy. Soc. (London)*, **A202**(1950), 301.
10. Alladi Ramakrishnan, *Proc. Cambridge Philos. Soc.*, **46**(1950), 595.
11. S. K. Srinivasan, *Stochastic Theory and Cascade Processes*, American Elsevier, New York, 1969, Chap. 3, Sec. 4.4.
12. S. K. Srinivasan, Point Processes and Product Densities (1970) (to be published).
13. N. Wiener, *Acta Math.*, **55**(1930), 117.
14. M. C. Wang and G. E. Uhlenbeck, *Rev. Mod. Phys.*, **17**(1945), 323.
15. H. R. Thompson, *Biometrika*, **42**(1955), 102.
16. D. R. Cox and W. L. Smith, *Biometrika*, **41**(1954), 91.
17. J. A. McFadden, *J. Roy. Statist. Soc.*, **B24**(1962), 364.
18. L. Janossy, *Proc. Roy. Irish Acad.*, **A53**(1950), 181.
19. J. de Boer, *Rept. Progr. Phys. (London)*, **12**(1949), 305.
20. D. R. Cox, *Renewal Theory*, Methuen, London, 1962.
21. S. K. Srinivasan, *J. Math. Phys. Sci.*, **1**(1967), 1.
22. M. S. Bartlett, *Stochastic Processes*, Cambridge University Press, 1955, p. 78.
23. W. Feller, *An Introduction to Probability Theory and its Applications*, John Wiley, New York, 1957, Chap. 17.
24. R. E. Bellman and T. E. Harris, *Proc. Nat. Acad. Sci. U.S.*, **34**(1948), 601.
25. R. E. Bellman, R. Kalaba, and G. M. Wing, *Proc. Nat. Acad. Sci. U.S.* **43**(1957), 517.
26. R. E. Bellman, R. Kalaba, and G. M. Wing, *J. Math. and Mech.*, **7**(1958), 149.
27. R. E. Bellman, R. Kalaba, and G. M. Wing, *J. Math. Phys.*, **1**(1960), 280.
28. J. E. Moyal, *J. Roy. Statist. Soc.*, **B11**(1949), 150.
29. Alladi Ramakrishnan and R. Vasudevan, *Proc. Indian Math. Soc.*, **24**(1960), 457.
30. Alladi Ramakrishnan, *Proc. Indian Acad. Sci.*, **A44**(1956), 428.

Chapter 2

RANDOM INTEGRATION

1. INTRODUCTION

The theory of integrals of deterministic functions and solution of differential equations, from the point of view of real analysis, is based on the fundamental notion of the limit of a sequence. With the advent of Lebesgue integration, measure theory, and generalized functions, it became possible to extend the idea of integration to the widest class of deterministic functions. In this context it is reasonable to investigate whether random functions can be treated in a similar manner. In fact, stochastic integrals which arise from very many physical situations were well known to electrical engineers as responses to random signals and noise. From a pure mathematician's point of view, limiting stochastic operations do not offer any difficulty, because they follow naturally from the notions of convergence almost everywhere, convergence in mean square and convergence in measure, provided we replace Lebesgue measure by probability measure and the space of reals by an appropriate function space. Although such a viewpoint is useful in that it provides a sound mathematical basis for the formulation of probability problems, it does not enable us to compute quantities of physical significance. Thus the situation is similar to the theory of Riemann integration, where the actual evaluation of integrals is achieved by the explicit use of primitives or of Simpson's formula rather than by the evaluation of the limit of the corresponding sequence. We shall use an exactly similar approach for the interpretation of integrals and show how many of the statistical properties of stochastic integrals can be deduced directly.

The layout of this chapter is as follows. Section 2 deals with the general method of arriving at the probability frequency function (p.f.f.) of weighted integrals of a class of random functions. Since the Poisson process plays a dominant role in a number of physical phenomena, we shall find it worthwhile to devote the Section 3 to present a general treatment of integrals associated with a Poisson point process and discuss its applications in different fields. The final section contains an account of integrals associated with general point processes.

2. INTEGRALS OF RANDOM FUNCTIONS

Many random processes occur in Physics and Engineering which can be symbolically represented by integrals or iterated integrals associated with a simple random process whose stochastic structure, including its mode of development, is known. One of the earliest examples in Electrical Engineering is the Schottky [1] effect (better known as the shot effect) arising from the fluctuations in the intensity of the stream of electrons flowing from the cathode to the anode. If $\phi(t)$ is the response at time t due to the arrival of an electron at the anode at time $t = 0$, the cumulative response $r(t)$ is given by

$$r(t) = \sum_i \phi(t - t_i), \tag{2.1}$$

where the t_i's are the random instants of arrivals of electrons. Equation (2.1) can be symbolically written

$$r(t) = \int_0^t \phi(t - t') \, dn(t'), \tag{2.2}$$

where $n(t)$ is the random variable representing the number of arrivals in the interval $(0, t)$. Campbell [2] obtained the first two moments of $r(t)$ for large values of t by performing an appropriate time average. Another example involving the integral of a random function is provided by the Brownian motion (see Einstein [3]). In this case the velocity u of a Brownian particle satisfies the equation of motion,

$$m\frac{du}{dt} = -fu + F(t), \tag{2.3}$$

where $-fu$ is the functional force due to the influence of the surrounding medium and $F(t)$ is a fluctuating force responsible for the random displacements suffered by the particle. Thus the velocity u can be expressed as a weighted integral of the random function $F(t)$.

With the advent of modern probability theory, many methods have been developed for the study of stochastic integrals. Uhlenbeck and Ornstein [4] were the first to study (2.3) and obtain the p.f.f. of u as well as the displacement. Much later, Doob [5] provided rigorous mathematical proofs of many of these results. An interesting account of all the connected results can be found in the survey article of Moyal [6] and in the classic monograph of Blanc-Lapierre and Fortet [7]. We deal with some aspects of these and other connected problems in Chapter 4. In this section we propose to describe a systematic method of treating such random integrals with special reference to their probability frequency functions, moments, and correlations.

Let us consider a Markovian stochastic process defined by the random variable $x(\tau)$, the process being homogenous with respect to τ. Thus the process is completely defined by the probability frequency function $\pi(x|x_0, t)$ where $\pi(x|x_0, t)\, dx$ denotes the probability that the random variable $x(\tau)$ takes a value between x and $x + dx$ at $\tau = t$ conditional on $x(0) = x_0$. In any experiment dealing with such a stochastic process progressing with respect to τ, we can plot the realized value of $x(\tau)$ in a certain interval $(0, t)$. If we denote by $x^R(\tau)$ the realized value of $x(\tau)$ corresponding to the parametric value τ, we note that in any physical realization (by experiment) it is possible to obtain a discrete set of values for $x(\tau_i)$, $0 \leqslant \tau_i \leqslant t$. The general problem of assigning a measure to a curve obtained by a set of values $x(\tau_i)$ is very difficult and will lead to pathological situations involving the joint distribution of the stochastic variable at a continuous infinity of points in $(0, t)$. However, for a certain class of problems it is possible to assign a probability measure to the path, and we will have occasion to demonstrate this possibility in Chapter 8 when we deal with path integrals. In Section 2.1 we present a method, due to Ramakrishnan [8], of dealing with the process when $x^R(\tau)$ is characterized by a finite number of transitions in any finite interval $(0, t)$, $x^R(\tau)$ remaining a constant between any two transitions. We then consider iterated integrals of such a random process. In Section 2.2 we present an alternative interpretation of such a special random process in terms of a weighted point process. Such a viewpoint enables us to study the p.f.f. of weighted integrals with the help of the characteristic functional of the basic point process.

2.1. The Basic Random Process and Its Integral

We assume that $x(t)$, besides being a Markovian and t-homogeneous process, is governed by the p.f.f. $\pi(x|x_0, t)$ having the property

$$\pi(x|x_0, \Delta) = R(x|x_0)\, \Delta + \delta(x - x_0)\left\{1 - \Delta \int_x R(x|x_0)\, dx\right\} + o(\Delta) \qquad (2.4)$$

for small values of Δ. From (2.4) we can derive the Kolmogorov forward differential equation satisfied by π. It is interesting to compare (2.4) with the Fokker–Planck approximation, where the behavior of $\pi(x|x_0, \Delta)$ is governed by the relation

$$\int_x \pi(x|x_0, \Delta)(x - x_0)^n\, dx = O(\Delta), \qquad n = 1, 2,$$

$$= o(\Delta), \qquad n > 2. \qquad (2.5)$$

The preceding condition will not be satisfied if (2.4) holds, for all the higher moments of $(x - x_0)$ vanish linearly in Δ. A detailed discussion of this point is presented in Chapter 4 when we deal with the Brownian motion. We shall assume also that

$$\int_{x'} R(x'|x)\, dx' < \infty.$$

This is an important condition which enables us to define a proper measure for the realized "trajectory" $x^R(t)$. The condition also permits us to conclude that the random variable undergoes a finite number of transitions in any finite interval $(0, t)$ with probability 1, the random variable remaining a constant between any two transitions. Defining

$$\alpha(x) = \int_{x'} R(x'|x)\, dx', \qquad (2.6)$$

we note that the probability measure for the typical trajectory $x^R(t)$ given by

$$x^R(t) = x_0 + (x_1 - x_0)H(t - t_1) + \ldots + (x_n - x_{n-1})H(t - t_n),$$
$$0 < t_1 < t_2 \ldots < t_n < t, \qquad (2.7)$$

can be written

$$e^{-\alpha(x_n)(t - t_n)} \prod_{i=0}^{n-1} e^{-\alpha(x_i)(t_{i+1} - t_i)} R(x_{i+1}|x_i)\, dx_{i+1}\, dt_{i+1},$$

where $t_0 = 0$ and $H(t)$ is the Heaviside unit function. Next it is easy to obtain $\pi(x|x_0, t)$ by the use of the expression above for the probability measure for a typical trajectory $x^R(\tau)$:

$$\pi(x|x_0, t) = \sum_{n=1}^{\infty} \int_0^t dt_n \int_0^{t_n} dt_{n-1} \ldots \int_0^{t_2} dt_1 \int_{x_{n-1}} \ldots \int_{x_1} \cdot$$
$$\left\{ e^{-\alpha(x)(t - t_n)} \prod_{i=0}^{n-1} e^{-\alpha(x_i)(t_{i+1} - t_i)} R(x_{i+1}|x_i)\, dx_{i+1} \right\} + e^{-\alpha(x_0)t} \delta(x - x_0), \quad (2.8)$$

where $x_n = x$. Equation (2.8) can also be obtained directly as a solution of the Kolmogorov forward differential equation. However, the use of (2.7) and the corresponding probability measure enables us to assign a measure for the integral of $x(t)$ over any given interval. If we consider the class of realized trajectories $x^R(t)$ and the integrals $\int_0^t x^R(t')\, dt'$ generated by them, we can conclude that the probability measure of

$$y^R(t) = \int_0^t x^R(t')\, dt' \qquad (2.9)$$

is the same as that of $x^R(t)$. Such a recourse provides a reasonable working definition of the integral of a basic random process. At first sight, this approach may appear to be restrictive from the point of view of stochastic convergence (see, for example, Ramakrishnan and Vasudevan [9]). However, there exist a variety of physical processes which can be represented as weighted integrals or iterated integrals of $x(t)$. This point is amply demonstrated in this as well as the succeeding chapters.

2.2. Iterated Integrals of the Basic Random Process

With the help of the basic random process, it is possible to generate a random process which does not apparently remain a constant between two transitions. This is achieved by defining a random variable $y_0(\tau)$ by

$$y_0(\tau) = \phi_0(\tau)x(\tau), \tag{2.10}$$

where $\phi_0(\tau)$ is a known deterministic function of τ, $x(\tau)$ being the B.R.P. introduced in Section 2.1. The realized value of the stochastic variable $y_0(\tau)$ is given by

$$y_0^R(\tau) = x_0\phi_0(\tau)H(\tau - t_1) + (x_1 - x_0)\phi_0(\tau)H(\tau - t_2) +$$
$$\ldots + (x_n - x_{n-1})\phi_0(\tau)H(\tau - t_n), \tag{2.11}$$

corresponding to the realized value of $x^R(\tau)$ given by (2.7). In a similar manner we can consider a general weighted integral of $x(\tau)$ given by

$$y_m(t) = \phi_m(t) \int_0^t y_{m-1}(\tau) \, d\tau, \tag{2.12}$$

$y_0(t)$ being defined by (2.10). The equation for $y_m^R(\tau)$, a typical realized trajectory of $y_m(\tau)$, is given by

$$y_m^R(\tau) = \phi_m(\tau) \int_0^\tau \phi_{m-1}(\tau_{m-1}) \, d\tau_{m-1} \int_0^{\tau_{m-1}} \phi_{m-2}(\tau_{m-2}) \, d\tau_{m-2} \cdots .$$
$$\int_0^{\tau_2} \phi_1(\tau_1) \, d\tau_1 \int_0^{\tau_1} \phi_0(\tau_0)x^R(\tau_0) \, d\tau_0. \tag{2.13}$$

The realized trajectories are continuous for all $m > 1$, the mth derivative of $y_m^R(\tau)$ having a finite number of finite discontinuities in the interval $(0, t)$. The probability measure of a typical trajectory corresponding to an interval $(0, t)$ is the same as that of the corresponding trajectory of the basic random process. Equation (2.12) shows that the p.f.f. of $y_m(t)$ cannot be obtained without reference to the random variable $y_{m-1}(t)$ and the p.f.f. of $y_{m-1}(t)$ cannot be obtained without any reference to the random variable $y_{m-2}(t)$, and so on. However, we can deal with the joint p.f.f. of the set $y_i(t)$, $i = 1, 2, \ldots, m$. If $\pi_m(y_1, y_2, \ldots, y_m, x|x_0, t)$ is the joint p.f.f. of the set $y_i(t)$ and x, then it is easy to show that π_m satisfies the equation (see Srinivasan [10])

$$\frac{\partial \pi_m(y_1, y_2, \ldots, y_m, x|x_0, t)}{\partial t} = - \alpha(x)\pi_m(y_1, y_2, \ldots, y_m, x|x_0, t) -$$
$$\sum_i \left[\frac{\phi_i'(t)}{\phi_i(t)} y_i + \phi_i(t)y_{i-1}(t) \right] \cdot \frac{\partial}{\partial y_i} \pi_m(y_1, y_2, \ldots, y_m, x|x_0, t) +$$
$$\int \pi_m(y_1, y_2, \ldots, y_m, x'|x_0)R(x|x') \, dx'. \tag{2.14}$$

Equation (2.14) is intractable in the general case. Explicit solution for $\pi_m(y_1, y_2, \ldots, y_m, x|x_0, t)$ is possible only in very special cases. If the basic random process is Poisson, π_m can be obtained explicitly (see, for example, Mathews and Srinivasan [11]).

We can avoid the joint p.f.f. of the set of random variables $y_i(t)$ by writing $y_m(t)$ as

$$y_m(t) = \int_0^t f_m(t, \tau)x(\tau)\, d\tau,$$

$$f_m(t, \tau) = \left[\phi_0(\tau) \int_\tau^t \phi_{m-1}(\tau_{m-1})\, d\tau_{m-1} \cdot \right.$$

$$\left. \int_\tau^{\tau_{m-1}} \phi_{m-2}(\tau_{m-2})\, d\tau_{m-2} \ldots \int_\tau^{\tau_2} \phi_1(\tau_1)\, d\tau_1 \right] \phi_m(t) \qquad (2.15)$$

and observing that the realized value of $y_m{}^R(t)$ corresponding to (2.7) is given by

$$y_m{}^R(t) = x_0 \int_0^t f_m(t, \tau)\, d\tau + (x_1 - x_0) \int_{t_1}^t f_m(t, \tau)\, d\tau +$$

$$\ldots + (x_n - x_{n-1}) \int_{t_n}^t f_m(t, \tau)\, d\tau. \qquad (2.16)$$

Thus the p.f.f. of $y_m(t)$ can be obtained by first writing the joint p.f.f. of $y_m(t)$ and $x(t_i)$, $i = 1, 2, \ldots, n$, and then integrating over t_i and summing over all values of n. The success of this technique when $x(t)$ is a point process having a simple structure has been demonstrated by Ramakrishnan and Srinivasan [12]. However, in the general case the integration and summation processes are tedious and do not lead to a simple closed expression for the p.f.f. of $y_m(t)$.

3. GENERALIZATION IN TERMS OF A POINT PROCESS

We have seen in Section 2.2 that the p.f.f. of integrals of a basic random process can be obtained only in some special cases. However, if $R(x|x')$ is of the special form given by

$$R(x|x')\, dx = R(x - x')\, d(x - x'), \qquad (3.1^*)$$

we can identify the basic random process with a suitable "weighted" point process (see, for example, Wold [13]). For in this case the typical realized trajectory corresponding to $x(\tau)$ is given by

$$x^R(\tau) = x_0 + \xi_1 H(\tau - t_1) + \xi_2 H(\tau - t_2) + \ldots + \xi_n H(\tau - t_n), \qquad (3.2)$$

* This particular form makes the forward differential equation satisfied by π_m amenable to Laplace transform technique in a more general case (see Srinivasan [10]).

where $\xi_1, \xi_2, \ldots, \xi_n$ are independent random variables (representing the jumps suffered by the random variable) identically distributed with the p.f.f. $\rho(\xi)$ given by

$$\rho(\xi) = \frac{R(\xi)}{\alpha}, \tag{3.3}$$

where α defined by (2.6) has now become a constant function in view of the assumption (3.1). Thus $x(t)$ is a point process defined on the t-axis (the points referring to the points at which the B.R.P. suffers jumps) with a random weight (or intensity) represented by $\xi(t_i)$, the $\xi(t_i)$'s being independently and identically distributed with the p.f.f. $\rho(\xi)$. However, the condition (3.1) is highly restrictive and in fact, taken along with the Markovian nature of the stochastic process represented by $x(t)$, converts the general point process into a Poisson process with the parameter α. To arrive at a more general process, we need not insist that the process be Markovian. Thus we can consider a t-homogeneous point process represented by the random variable $dN(t)$ (see, for example, reference [11]), where $N(t)$ is the random variable representing the total number of points realized in the interval $(0, t)$. Then $x(t)$ can be conveniently represented as

$$x(t) = \int_0^t dN(\tau)\xi(\tau) \tag{3.4}$$

where $\{\xi(\tau), \tau \in (0, t)\}$ represents a set of random variables which are independently and identically distributed with the p.f.f. $\rho(\xi)$. In this case $\rho(\xi)$ need not be related to $R(\xi)$ (defined by (3.3)).

Next we observe that a typical realized trajectory of $x(\tau)$ corresponding to the interval $(0, t)$ is still described by (3.2). However, the probability measure corresponding to such a realization is given by

$$P(t_1, t_2, \ldots, t_n, t) \, dt_1 \, dt_2 \ldots dt_n \, \rho(\xi_1)\rho(\xi_2) \ldots \rho(\xi_n) \, d\xi_1 \, d\xi_2 \ldots, d\xi_n,$$

where $P(t_1, t_2, \ldots, t_n, t) \, dt_1 \, dt_2 \ldots dt_n$ denotes the probability that the ith point of the point process counted from $\tau = 0$ occurs between t_i and $t_i + dt_i$, $i = 1, 2, \ldots, n$, and that no further point occurs in the interval (t_n, t). Since all the statistical properties of the basic point process are assumed to be given, the function P can be determined. Because a point process is uniquely determined by the characteristic functional, we shall assume that this is known. Let $\Phi([\theta])$ be the characteristic functional of the point process represented by the random variable $dN(t)$. Our objective is to demonstrate the possibility of obtaining the statistical characteristics of $y_m(t)$ defined by (2.15) with the help of $\Phi([\theta])$.

Defining $C_m([\theta])$, the characteristic functional of $y_m(t)$, by

$$C_m([\theta]) = \mathbf{E}\left\{\exp i \int_0^\infty \theta(t) y_m(t)\, dt\right\}, \qquad (3.5)$$

we note that $C_m([\theta])$ can be directly expressed in terms of an integral involving $dN(\tau)$. Substituting from (2.15) and making use of (3.3), we find

$$C_m([\theta]) = \mathbf{E}\left\{\exp i \int_0^\infty \xi(\tau)\, dN(\tau)\psi_m(\tau)\right\}, \qquad (3.6)$$

where

$$\psi_m(\tau) = \int_\tau^\infty \theta(t)\, dt \int_\tau^t f_m(t, t')\, dt'. \qquad (3.7)$$

Next we evaluate the expectation implied on the right-hand side of equation (3.6). Replacing the integral by the sum, we find

$$C_m([\theta]) = \mathbf{E}\{\exp i \sum_j \xi(\tau_j)\, dN(\tau_j)\psi_m(\tau_j)\, \Delta\tau_j\}$$

$$= \mathbf{E}\{\prod_j [\exp i\xi(\tau_j)\psi_m(\tau_j)\, \Delta\tau_j]^{dN(\tau_j)}\}. \qquad (3.8)$$

If we observe that the random variable $dN(\tau_j)$ effectively takes the values 0 and 1 (see, for example, Ramakrishnan [14]) and that the random variables $\xi(\tau_j)$ are independently and identically distributed, we find

$$C_m([\theta]) = \Phi([\chi_m]), \qquad (3.9)$$

where χ_m is a functional of θ given by

$$\chi_m(\tau) = \frac{1}{i} \ln \int_{-\infty}^{+\infty} e^{i\psi_m(\tau)\xi} \rho(\xi)\, d\xi. \qquad (3.10)$$

Equations (3.9) and (3.10) constitute the explicit solution of our problem. In particular, the characteristic functional of the process $x(t)$ can be obtained by replacing $\psi_m(\tau)$ by $\int_\tau^\infty \theta(t)\, dt$. Equation (3.9) is valid even in a more general nonstationary situation when $\rho(\xi)$ is also a function of τ.

A special case of interest arises if $\xi(\tau)$ equals a constant (which can be taken to be unity) with probability 1. Then $x(t)$ represents the number of points in the interval $(0, t)$. Ramakrishnan [15, 16] studied this problem in detail and obtained the moments of $y_m(t)$ as weighted integrals of the product densities (see [14]) corresponding to the point process generated by the random variable $dN(t)$. All the results obtained by Ramakrishnan can be deduced from (3.9) by taking appropriate functional derivatives of $C_m([\theta])$. In addition, we can also obtain the correlation of $y_m(t)$ of any order.

3

4. WEIGHTED INTEGRALS OF SOME SPECIAL POINT PROCESSES

Although the results presented in Section 3 constitute the complete solution, it is worthwhile to consider special cases of point processes comprising the Poisson process, the Yule–Furry process, and the birth and death process, as they are of great practical significance in Physics and Engineering. As a great deal of work has been done in this direction, we propose to give a short survey of all the efforts in obtaining the statistical properties such as the p.f.f., moments, and correlations.

As mentioned in Section 2, integrals associated with a Poisson process were first investigated by Campbell [2], who in 1905 gave a probabilistic account of the fluctuation phenomena encountered in the emission of rays by a radioactive substance with special reference to Geiger's experiment. A lucid account of the method of performing time averages and arriving at the moments and p.f.f. of the response function defined by (2.1) will be found in the survey article of Rice [17]. The stochastic integrals associated with the Poisson process were investigated in great detail by Blanc-Lapierre and Fortet [7], who obtained explicitly the characteristic function of the integrals of the type $y_m(t)$ defined by (2.15). The purpose of this section is to project some of the recent developments including some results on nonstationary processes.

4.1. Poisson Process: Markovian Description and Equivalent Processes

Let us consider the random variable $y_m(t)$ defined by (2.12) when $x(t)$ is the random variable representing the number of Poisson events that have occurred in the interval $(0, t)$. It is convenient to denote $x(t)$ by $n(t)$ to bring out the discrete nature of the random variable. We observe that, although the process $n(t)$ is Markovian, $y_m(t)$ is no longer so. Hence the familiar method of increasing t by Δ and relating the p.f.f. of y_m at $t + \Delta$ in terms of the p.f.f. of y_m at t fails. However, we can consider the set of random variables $y_i(t)$, $i = 1, 2, \ldots, m$, and $n(t)$. The resulting vector process is Markovian in nature, and we can readily write the Kolmogorov forward differential equation for the joint p.f.f. of the set of random variables constituting the vector process. This was done by Mathews and Srinivasan [11], who obtained in an explicit form the Laplace transform of the p.f.f. when the functions $\phi_i(t)$, $i = 0, 1, 2, \ldots, m$, are positive definite. We can avoid the joint p.f.f. of the set of variables $y_i(t)$ and directly deal with the joint p.f.f. of $y_m(t)$ and $x(t_i)$. If we notice that $y_m(t)$ is given by

$$y_m(t) = \int_0^t dn(t')F_m(t, t'), \qquad (4.1)$$

where

$$F_m(t, t') = \int_{t'}^{t} f_m(t, t'') \, dt'', \tag{4.2}$$

then the random variable is easily identified with a sum obtained by associating $F_m(t, t_i)$ with each Poisson event occurring between t_i and $t_i + dt_i$. The probability measure for the occurrence of an event between t_i and $t_i + dt_i$, $i = 1, 2, \ldots, n$, there being no further events in $(0, t)$ can be readily written. From this it is possible to obtain the p.f.f. of $y_m(t)$ explicitly by integrating over the variables $t_1, t_2, \ldots, t_n$ (see Srinivasan [18]).

We wish to describe another method due to Ramakrishnan [8] which completely dispenses with the joint p.f.f. of $y_m(t)$ and $x(t_i)$ or $y_i(t)$ and $x(t)$. This method is successful in the special case corresponding to $\phi_i(t) = 1$, $i = 1, 2, \ldots, n$. In that case, $y(t)$ is given by

$$y_m(t) = \int_{0}^{t} n(\tau) \frac{(t - \tau)^m}{m!} \, d\tau, \tag{4.3}$$

where $n(\tau)$ is assumed to be a homogeneous Poisson process. Ramakrishnan [8] noticed that the random integral given by

$$y_m^*(t) = \int_{0}^{t} n(t - \tau) \frac{\tau^m}{m!} \, d\tau \tag{4.4}$$

has the same p.f.f. as $y_m(t)$ in view of the symmetric nature of the Poisson process. It should be noted that, as τ is increased from 0 to t, the realized trajectory, $n^R(t - \tau)$, is different from the $n^R(\tau)$ arising from the original Poisson process. However, the probability measure for the integral is the same in either case because of the independent nature of the Poisson events. Thus the random process $y_m^*(t)$ can be considered "equivalent" to the random process $y_m(t)$. The p.f.f. of $y_m^*(t)$ is easily obtained without recourse to the joint p.f.f. of the set $y_i^*(t)$, $i = 1, 2, \ldots, n$. The notion of equivalence extends only up to the equality of the p.f.f. of $y_m(t)$ and $y_m^*(t)$ for any given t. In fact, if we ask more detailed questions relating to the correlations of $y_m(t)$ at two or more points on the t-axis, the equivalence breaks down. In fact, this was demonstrated for y_2 by Ramakrishnan and Srinivasan [19] (see also Srinivasan [18]) in their investigations relating to some correlation problems in the brightness of the Milky Way.

There is another technique due to Ramakrishnan [16] which consists in defining the random variable $z_m(a, t)$ by

$$z_m(a, t) = \int_{0}^{t} F_m(a, \tau) \, dn(\tau), \tag{4.5}$$

where a is any general parameter. Since $z_m(a, t) = y_m(t)$ when $a = t$, the p.f.f.s of $y_m(t)$ and $z_m(a, t)$ are equal at the instant $t = a$. However, the evolution of the two processes are quite different. In fact, $z_m(a, t)$ is a basic random process, whereas $y_m(t)$ is a continuous random variable whose relationship to a basic random process is nontrivial. Although this method is useful in the determination of the p.f.f. at any particular t and is applicable even when $n(t)$ is not a Poisson process, more detailed questions like the correlations of y_m at different t's cannot be answered. We do not propose to go into greater detail since all these results can be obtained from the characteristic functional of $x(t)$, as demonstrated in Section 3.

4.2. Poisson Process: Solution by the Use of the Characteristic Functional

The characteristic functional of $y_m(t)$ can be obtained from (3.9) by appropriate substitution for the functional χ_m. In this section we present an alternative derivation which brings out clearly the independent nature of the individual Poisson events. The functional $C_m([\theta])$ as defined by (3.6) can be written

$$C_m([\theta]) = E\{\prod_j [\exp i\xi(\tau_j)\psi_m(\tau_j)\, dN(\tau_j)\, \Delta\tau_j]\}. \tag{4.6}$$

Next we observe that

$$\text{Prob}\{dN(\tau_j) = 1\} = \lambda\, \Delta\tau_j + o(\Delta\tau_j),$$
$$\text{Prob}\{dN(\tau_j) = 0\} = 1 - \lambda\, \Delta\tau_j + o(\Delta\tau_j), \tag{4.7}$$

and that $dN(\tau_j)$ for different distinct τ_j's are independently and identically distributed. Thus (4.6) can be written

$$C_m([\theta]) = E_\xi\{\prod_j [(1 - \lambda\, \Delta\tau_j) + e^{i\xi(\tau_j)\psi_m(\tau_j)}\, \lambda\, \Delta\tau_j])\}, \tag{4.8}$$

where E_ξ denotes the expectation value with respect to ξ. Performing this average, we obtain

$$C_m([\theta]) = \exp\left[-\int_0^\infty \lambda\{1 - g(\psi_m(t))\}\, dt \right], \tag{4.9}$$

where

$$g(u) = \int_{-\infty}^{+\infty} e^{i\xi u}\, \rho(\xi)\, d\xi. \tag{4.10}$$

This result is equally valid in the case of an inhomogeneous Poisson process if we replace λ by $\lambda(t)$ in the integrand on the right-hand side of (4.9). The characteristic functional in the explicit form above constitutes the complete solution of the problem of shot noise as well as the response of general vibratory systems to Poisson excitations. If we consider the special case when ξ equals a constant (equal to unity) with probability 1. Eq. (4.9) takes the simple form

$$C_m([\theta]) = \exp\left[- \int_0^\infty \lambda(1 - \exp i\psi_m(\tau))\, d\tau \right]. \tag{4.11}$$

We can specialize still further by choosing

$$\theta(\tau) = \theta_0\, \delta(\tau - t), \tag{4.12}$$

in which case we can identify $C_m([\theta])$ with the characteristic function of the random variable $y_m(\tau)$. Thus we have

$$C([\theta]) = \mathbf{E}\{\exp i\theta_0 y_m(t)\}$$
$$= \exp\left[\int_0^t \lambda(1 - e^{i\theta_0 F_m(t,\tau)})\, d\tau \right]. \tag{4.13}$$

A special case of (4.13) has been obtained by Mathews and Srinivasan [11] when the functions $\phi_i(t)$ are positive definite. Ramakrishnan [16] derived (4.13) by considering the random variable $z_m(a, t)$ as defined by (4.5) and recognizing it as equivalent to the process defined by $y_m(t)$.

A result which is more general than (4.13) can be obtained from (4.9) by taking $x(t)$ to be a weighted Poisson process, the weights at different t's being independently and identically distributed. The characteristic function of $y_m(t)$ in this case is given by

$$C_m(\theta_0) = \exp\left[- \lambda \int_0^t (1 - g(\theta_0 f_m(t, \tau))\, d\tau \right]. \tag{4.14}$$

This result has been obtained by Rice [17] when $f_m(t, \tau)$ is a function of $(t - \tau)$ only.

4.3. The Symmetric Oscillatory Poisson Process

Let events occur along t-axis in a Poisson manner and let us associate with the ith event a random variable which take the values $+1$ and -1 with probability p and $1 - p$, respectively. Then we can define the random variable $x(t)$,

$$x(t) = \sum_i \xi_i \tag{4.15}$$

which can assume integral values. Thus $x(t)$ is a basic random process, the typical trajectory of $x(t)$ consisting of lines parallel to the t-axis with jumps of unit magnitude (positive and negative). Such a random variable was studied by Ramakrishnan and Mathews [20], who dealt with the p.f.f. of $x(t)$ and its integral in connection with some models for multiple scattering in multiplicative processes.

The characteristic functional of $y_m(t)$ can be obtained from (3.9) if we set

$$\rho(\xi) = p\, \delta(\xi - 1) + (1 - p)\, \delta(\xi + 1). \tag{4.16}$$

Thus $C([\theta])$ is given by

$$C([\theta]) = \exp - \int_0^\infty \lambda\{1 - p \exp i\psi_m(\tau) - (1 - p) \exp - i\psi_m(\tau)\}\, d\tau. \tag{4.17}$$

We can still further specialize by choosing

$$\theta(\tau) = \theta_0\, \delta(\tau - t), \tag{4.12}$$

which leads to the identification of $C_m[\theta]$ with the characteristic function of $y_m(\theta)$:

$$C_m(\theta_0) = \mathbf{E}\{\exp i\theta_0 y_m(t)\}$$

$$= \exp\left[\int_0^t \lambda(1 - p \exp i\theta_0 f_m(t, \tau) -\right.$$

$$\left. (1 - p) \exp - i\theta_0 f_m(t, \tau))\, d\tau\right]. \tag{4.18}$$

By taking appropriate forms of $f_m(t, \tau)$ we can recover all the results obtained by Ramakrishnan and Mathews [20]).

4.4. Integrals of Birth and Death Processes

Birth and death processes have been studied in the past in great detail (see, for example, Kendall [21, 22], Harris [23], and Srinivasan [24]). If $x(t)$ is taken to be the population size at time t in a simple homogeneous birth and death process, an integral of the type $\int_0^t x(t')\, dt'$ acquires a special significance because it provides a measure of the amount of toxin produced in time $(0, t)$ by a colony of bacteria. The p.f.f. of the integral of $x(t)$ was studied by Puri [25] by considering the joint p.f.f. of $x(t)$ and $y(t)$ defined by

$$y(t) = \int_0^t x(t')\, dt'. \tag{4.19}$$

Using the method outlined in Section 2.2, the joint characteristic function of $x(t)$ and $y(t)$, defined by

$$\Psi(u, v, t) = \mathbf{E}\{\exp[iux(t) + ivy(t)]\}, \tag{4.20}$$

can be shown to satisfy the differential equation λ and μ are the birth and death rates.

$$\frac{\partial \Psi(u, v, t)}{\partial t} = \frac{\partial \Psi(u, v, t)}{\partial u}[v + i\{(\mu + \lambda) - \mu e^{-iu} - \lambda e^{iu}\}], \tag{4.21}$$

Assuming that the population size is m at $t = 0$, Puri [25] obtained the following explicit expression for $\Psi(u, v, t)$:

$$\Psi(u, v, t) = \left[r_2 + \frac{r_1 - r_2}{1 - e^{\lambda(r_1 - r_2)t}(e^{iu} - r_1)/(e^{iu} - r_2)} \right]^m, \tag{4.22}$$

where r_1 and r_2 are the roots of the equation

$$\lambda z^2 + (iv - \mu - \lambda)z + \mu = 0. \tag{4.23}$$

Peter Jagers [26] has considered the integral of $x(t)$ when the parameters λ and μ are age-dependent. Confining himself to the moments of $y(t)$, Peter Jagers has dealt with some of the limiting properties of the moments.

We observe next that the properties of any weighted integral of the birth and death process can be obtained if we are in possession of the characteristic functional $\Phi(\theta, t)$ defined by

$$\Phi(\theta, t) = \mathbf{E}\left\{ \exp i \int_0^t dx(t')\theta(t') \right\}. \tag{4.24}$$

However, it is very difficult to obtain an explicit expression for $\Phi(\theta, t)$. Rangan* has shown by using the method outlined in Section 2.2 that in the case of constant λ and μ, the functional $G(u, \theta, t)$ defined by

$$G(u, \theta, t) = \mathbf{E}\left\{ u^{x(t)} \exp i \int_0^t \theta(t') \, dx(t') \right\} \tag{4.25}$$

satisfies the equation

$$\frac{\partial G(u, \theta, t)}{\partial t} = \frac{\partial G(u, \theta, t)}{\partial u}(1 - ue^{i\theta(t)})(\lambda u - \mu e^{-i\theta(t)}). \tag{4.26}$$

It is indeed difficult to obtain the explicit solution of (4.26).

We finally observe that the characteristic functional of the type defined by (4.24) can be obtained for a pure birth process (Yule–Furry process) when λ is a function of time only (see Srinivasan [27]). For in this case the probability that n births occur in the time intervals $(t_1, t_1 + dt_1)$, $(t_2, t_2 + dt_2)$, $\ldots$, $(t_n, t_n + dt_n)$ with no births anywhere else in the time interval $(0, t)$ is given by

$$e^{-\Lambda(t_1)}\lambda(t_1) \, dt_1 \exp\{-2[\Lambda(t_2) - \Lambda(t_1)]\}2\lambda(t_2) \, dt_2 \ldots$$
$$\exp\{-n[\Lambda(t_n) - \Lambda(t_{n-1})]\}n\lambda(t_n) \, dt_n \exp\{-(n+1)[\Lambda(t) - \Lambda(t_n)]\}, \tag{4.27}$$

where

$$\Lambda(t) = \int_0^t \lambda(t') \, dt'. \tag{4.28}$$

* Private communication.

Using the expression (4.27), Srinivasan [27] obtained the following expression for the characteristic functional:

$$\Phi(\theta, t) = \left[1 + \int_0^t \lambda(u) \, e^{\Lambda(u)} (1 - e^{i\theta(u)}) \, du \right]^{-1}. \tag{4.29}$$

As we have already mentioned, weighted integrals of the Yule–Furry process have been studied by Ramakrishnan and Srinivasan [12]. By taking appropriate forms of θ we can recover all the results obtained in [12].

REFERENCES

1. W. Schottky, *Ann. d. Physik.*, **57**(1918), 541.
2. N. Campbell, *Proc. Cambridge Philos. Soc.*, **15**(1909), 117.
3. A. Einstein, *Ann. d. Physik*, **17**(1905), 549; **19**(1906), 371.
4. G. E. Uhlenbeck and L. S. Ornstein, *Phys. Rev.*, **36**(1930), 823.
5. J. L. Doob, *Ann. Math.*, **43**(1942), 351.
6. J. E. Moyal, *J. Roy. Statist. Soc.*, **B11**(1949), 150.
7. A. Blanc-Lapierre and R. Fortet, *Théorie des Fonctions Aléatoires*, Masson et Cie., Paris, 1953.
8. Alladi Ramakrishnan, *Proc. Koninkl. Nederl. Akad. Sci. Wetenschap.*, **58**(1955), 470, 634.
9. Alladi Ramakrishnan and R. Vasudevan, *Proc. Indian Math. Soc.*, **24**(1960), 457.
10. S. K. Srinivasan, *Z. Angew. Math. Mech.*, **43**(1963), 259.
11. P. M. Mathews and S. K. Srinivasan, *Proc. Nat. Inst. Sci. (India)*, **A22**(1956), 369.
12. Alladi Ramakrishnan and S. K. Srinivasan, *Pub. Inst. Statist.*, **5**(1956), 95.
13. H. Wold, *Skand. Aktuarietidskr*, **31**(1948), 229.
14. Alladi Ramakrishnan, *Proc. Cambridge Philos. Soc.*, **46**(1950), 595.
15. Alladi Ramakrishnan, *Proc. Cambridge Philos. Soc.*, **49**(1953), 473.
16. Alladi Ramakrishnan, *Proc. Koninkl. Nederl. Akad. Wetenschap.*, **A59**(1956), 120.
17. S. O. Rice, *Bell Sys. Tech. J.*, **23**(1944), 282; **25**(1945), 46.
18. S. K. Srinivasan, Some Physical Applications of Random Functions and Their Associated Integrals, Ph.D. Thesis, Madras University, 1957.
19. Alladi Ramakrishnan and S. K. Srinivasan, *Astrophys. J.*, **123**(1956), 479.
20. Alladi Ramakrishnan and P. M. Mathews, *Proc. Indian Acad. Sci.*, **A43**(1956), 84.
21. D. G. Kendall, *Ann. Math. Statistics*, **19**(1948), 1.
22. D. G. Kendall, *J. Roy. Statist. Soc.*, **B12**(1950), 278.
23. T. E. Harris, *The Theory of Branching Processes*, Springer-Verlag, Berlin, 1963.
24. S. K. Srinivasan, *Stochastic Theory and Cascade Processes*, American Elsevier, New York, 1969, Chap. 9.
25. P. S. Puri, *Biometrika*, **53**(1966), 61.
26. Peter Jagers, *Biometrika*, **54**(1967), 263.
27. S. K. Srinivasan, Point Processes and Product Densities (to be published).

Chapter 3

ORDINARY DIFFERENTIAL EQUATIONS CONTAINING RANDOM FUNCTIONS

1. INTRODUCTION

One of the main fields of the application of Chapter 2 is the solution of differential equations involving random functions. In fact, many processes occur in Physics and Engineering which can be represented by a differential equation of the form

$$Ly(t) = x(t), \tag{1.1}$$

where L is a differential operator and $x(t)$ is a random function. Although such random equations have been studied and developed by many workers in different disciplines of Physics and Engineering, only recently have attempts been made to develop the theory of random equations (see, for example, Bharucha-Reid [1, 2]). In fact, the field has not been sufficiently explored, and almost all the results that have been obtained so far refer to very special situations. In this chapter we propose to demonstrate the utility of the methods developed in Chapter 2 by considering a particular class of random differential equations that are amenable to explicit treatment.

The organization of the present chapter is as follows. Section 2 deals with the general method of solution of linear equations with random forcing functions. Some relevant physical examples are discussed in detail. In the next section we deal with general linear stochastic differential equations where the stochastic nature arises from the coefficients in addition to the forcing terms. We then discuss general differential equations with random initial (or boundary) conditions.

2. LINEAR SYSTEMS SUBJECT TO RANDOM EXCITATIONS

Let us consider the differential equation of mth order:

$$\frac{d^m y}{dt^m} + a_1 \frac{d^{m-1} y}{dt^{m-1}} + \ldots + a_m y = k(t) x(t), \tag{2.1}$$

where $a_1, a_2, \ldots, a_m$ and $k(t)$ are fully determinate functions of t, and x is a random function of t. The problem of studying the probability distribution of y given the distribution of x is indeed a difficult one. However, we can deal with the problem quite satisfactorily if we are able to obtain the solution of (2.1) as

$$y(t) = \phi_m(t) \int_0^t \phi_{m-1}(t_{m-1})\, dt_{m-1} \int_0^{t_{m-1}} \phi_{m-2}(t_{m-2})\, dt_{m-2} \cdots$$
$$\cdot \int_0^{t_1} \phi(t_0)x(t_0)\, dt_0. \qquad (2.2)$$

The passage from (2.1) to (2.2) for any general m is, by itself, a difficult problem except in the special case where the a_i's are constants. It is interesting to note that the solutions of a number of equations of type (2.1) representing physical processes are capable of being expressed in the form (2.2). In Sections 2.1 and 2.2 we outline the method.

2.1. Linear Differential Equations of First and Second Order*

Let us consider the random variable $y(t)$ defined by

$$\frac{dy}{dt} + a(t)y = k(t)x(t). \qquad (2.3)$$

Our objective is to obtain the probability frequency function of $y(t)$ given the statistical properties of $x(t)$. The first step consists in formally writing the solution of (2.3) as

$$y(t) - y_0 e^{-b(t)} = y'(t) = \int_0^t k(\tau)x(\tau)e^{-b(t)-b(\tau)}\, d\tau, \qquad (2.4)$$

where

$$b(t) = \int_0^t a(t')\, dt', \qquad (2.5)$$

as if $x(t)$ were a fully determinate function. A complete justification for (2.4) can be provided by observing that, corresponding to any particular realized trajectory $x^R(\tau)$ of the random variable $x(\tau)$, the solution of (2.3) can be written in the form (2.4) and the initial condition, defined as $y(t) = y_0$ at $t = 0$, implies that for all possible realizations, y takes the value y_0 at $t = 0$. The modifications needed are trivial if y_0 is governed by a probability frequency function.

It is clear that the statistical properties of $y(t)$ or $y(t) - y_0 e^{-bt}$ can be studied as a special case of the random integral given by Eq. (2.15) of Chapter 2 provided we take

$$f_m(t, \tau) = k(\tau) \exp - \{b(t) - b(\tau)\}. \qquad (2.6)$$

If we define $z(t)$ by

$$z(t) = y(t)e^{b(t)} - y_0 = \int_0^t k(\tau)e^{+b(\tau)}x(\tau)\, d\tau, \qquad (2.7)$$

* In Section 2.1 we present the results of Mathews and Srinivasan [3].

it is easy to arrive at the forward differential equation

$$\frac{\partial \pi(z, x, t)}{\partial t} = -\alpha(x)\pi(z, x, t) + \int \pi(z, x', t)R(x|x')\,dx' -$$

$$xk(t)e^{b(t)}\frac{\partial \pi(z, x, t)}{\partial z} \qquad (2.8)$$

governing $\pi(z, x, t)$, the joint probability frequency function of z and x when $x(t)$ represents a basic random process (see Section 2.1 of Chapter 2). The initial condition is given by

$$\pi(z, x, 0) = \delta(z)\,\delta(x - x_0), \qquad (2.9)$$

where $\delta(x)$ is the Dirac delta function.

On the other hand, if $x(t)$ represents a weighted point process defined by Eq. 3.4 of Chapter 2, the characteristic functional of z or y can be readily obtained.

Next we discuss the general second-order linear differential equation

$$\frac{d^2y}{dt^2} + a_1(t)\frac{dy}{dt} + a_2(t)y = k(t)x(t). \qquad (2.10)$$

If we make the transformation*

$$y = we^{-\frac{1}{2}b_1(t)}, \qquad (2.11)$$

where

$$b_1(t) = \int_0^t a_1(t')\,dt',$$

(2.10) above can be written

$$\frac{d^2w}{dt^2} + p(t)w = k(t)x(t), \qquad (2.12)$$

where

$$p(t) = -\tfrac{1}{4}[a_1(t)]^2 - \tfrac{1}{2}\frac{d}{dt}a_1(t) + a_2(t). \qquad (2.13)$$

Next we use the identity

$$\left\{\frac{d}{dt} - q(t)\right\}\left\{\frac{d}{dt} + q(t)\right\}w \equiv \frac{d^2w}{dt^2} + w\left\{\frac{dq(t)}{dt} - [q(t)]^2\right\} \qquad (2.14)$$

to observe that, if $w(t)$ is a solution of the Riccati equation

$$\frac{dq(t)}{dt} - [q(t)]^2 = p(t), \qquad (2.15)$$

---

* This is a well-known standard method of reducing a second-order differential equation. See, for example, Whittaker and Watson [4].

then (2.12) can be written in the form

$$\left\{\frac{d}{dt} - q(t)\right\}\left\{\frac{d}{dt} + q(t)\right\}w(t) = k(t)x(t).$$ (2.16)

Thus the solution can be written

$$w(t) = \int_0^t \exp\left\{-r(t) + r(\tau)\right\} d\tau \int_0^\tau \exp\left\{r(\tau) - r(\tau')\right\}k(\tau')x(\tau')\,d\tau',$$ (2.17)

where

$$r(t) = \int_0^t q(t')\,dt'$$ (2.18)

for zero initial conditions. Equation (2.17) expresses $y(t)$ effectively as an iterated integral. Thus, if $x(t)$ is a basic random process, the joint probability frequency function of y and x satisfies an equation similar to (2.8). On the other hand, if $x(t)$ were a weighted point process, the characteristic function of $y(t)$ is directly expressible in terms of the characteristic functional of the point process generating $x(t)$.

2.2. Higher-Order Linear Differential Equations

In a similar manner, we can treat higher-order differential equations. For instance, we can write (1.1) in the vector form

$$\frac{d\mathbf{y}(t)}{dt} + A(t)\mathbf{y}(t) = \mathbf{x}(t),$$ (2.19)

where $A(t)$ is a general $n \times n$ matrix and $\mathbf{x}(t)$ is a vector, the components of which represent a set of statistically independent basic random processes. There is no general method of arriving at the explicit solution of an equation of the type (2.19). However, we can recall a theorem (see, for example, Coddington and Levinson [5]) which states that, if Φ is a fundamental matrix satisfying the homogeneous equation corresponding to (2.19), then the solution of (2.19) for zero initial conditions is given by

$$\mathbf{y}(t) = \Phi(t) \int_0^t \Phi^{-1}(\tau)\mathbf{x}(\tau)\,d\tau.$$ (2.20)

Equation (2.20) is useful in finding the moments and correlations of $y(t)$. For instance, the correlation of $\mathbf{y}$ of degree 2 is obtained by taking the average of the direct product of both sides of (2.20) corresponding to two different values of t. On the other hand, we can deal with (2.19) directly and obtain the probability frequency function of the set y_i and x_i by writing a Kolmogorov forward differential equation similar to (2.8).

If, in (2.19), $A(t)$ is a constant matrix, it is quite easy to arrive at the statistical properties of $\mathbf{y}(t)$. An analysis running almost parallel to the development of the theory of deterministic linear system was carried out by Srinivasan

[6]. We present here some of the salient features. If $y_i(t)$ is any component of $\mathbf{y}(t)$, $y_i(t)$ can be explicitly written in the form

$$y_i(t) = \int_0^t f_{ij}(t - \tau)x_j(\tau)\,d\tau. \tag{2.21}$$

If $x_j(t)$ for each j is a basic random process, then (2.21) is a linear combination of integrals of the type discussed in Chapter 2. If, on the other hand, $\{x_j(t)\}$ constitutes a set of independent weighted point processes, the characteristic functional of y_i can be expressed in terms of the characteristic functional of the individual $x_j(t)$. We illustrate this in Section 2.3 when we deal with the oscillations of galvanometers subject to a random couple.

Finally, we observe that it is possible with (2.19) by converting it into an mth-order differential equation of the form

$$\frac{d^m y}{dt^m} + p_1(t)\frac{d^{m-1}y}{dt^{m-1}} + \ldots + p_m y = F(t), \tag{2.22}$$

where $F(t)$ represents a linear combination of the basic random processes. In this case we can use the theorem of Schleisinger [7] (see also Polya [8]) to write (2.22) in the form

$$\frac{d}{dt}\left[g_1(t)\frac{d}{dt}\left[g_2(t)\frac{d}{dt}\left[\ldots\frac{d}{dt}[g_m(t)y]\ldots\right]\right]\right] = F(t), \tag{2.23}$$

an equation which enables us to put $y(t)$ as a weighted integral of $F(t)$ of the form

$$y(t) = \int_0^t h(t,\tau)F(\tau)\,d\tau,$$

the advantage of such a form being apparent. However, the passage from (2.22) to (2.23) will itself involve the solution of an mth-order differential equation of a different type which by itself may be difficult to achieve.

2.3. Physical Applications

An example of a process which leads to an ordinary differential equation of the first order of type (2.3) is provided by the fluctuating voltage at the anode of a thermionic valve. The fluctuations considered here are due to the discrete nature of the electric charge. When the average current flowing through a thermionic valve remains constant, the arrival times of electrons at the anode can reasonably be assumed to have a Poisson distribution. If the circuit between the anode and earth can be considered equivalent to a resistance R and a capacity C in parallel, the voltage drop $V(t)$ between battery and the anode satisfies the equation

$$\frac{dV}{dt} + \frac{V}{RC} = -\frac{\varepsilon}{C}\frac{dn(t)}{dt}, \tag{2.25}$$

where $n(t)$ represents the number of electrons that have arrived in $(0, t)$. Thus the solution (2.25) is given by

$$V' = V - V_0 e^{-t/RC} = -\frac{\varepsilon}{C} \int_0^t e^{-(t-\tau)/RC} \, dn(\tau). \qquad (2.26)$$

The characteristic functional of $V'(t)$ can be obtained from Eq. (4.11) of Chapter 2:

$$C([\theta]) = \mathbf{E}\left\{\exp i \int_0^\infty \theta(t) V'(t) \, dt\right\}$$

$$= \exp -\left[n_0 \int_0^\infty (1 - \exp i\psi(t)) \, dt\right], \qquad (2.27)$$

where

$$\psi(t) = -\frac{\varepsilon}{C} \int_t^\infty \theta(\tau) e^{-(t-\tau)/RC} \, d\tau. \qquad (2.28)$$

On the other hand, the characteristic function of $V'(t)$ is given by

$$C(\theta_0) = \mathbf{E}\{\exp i\theta_0 V'(t)\}$$

$$= \exp -n_0 \int_0^t \left\{1 - \exp\left[-\frac{i\theta_0 \varepsilon}{C} \exp\left\{-\frac{t - t'}{RC}\right\}\right]\right\} dt, \qquad (2.29)$$

where n_0 is the average number of electrons reaching the anode per unit time (so that $n_0 \varepsilon$ is the magnitude of the average electric current). The moments of V' are given by

$$\mathbf{E}\{V'(t)\} = -n_0 \varepsilon R(1 - e^{-t/RC}), \qquad (2.30)$$

$$\mathbf{E}\{[V'(t)]^2\} = n_0^2 \varepsilon^2 (1 - e^{-t/RC})^2 + \frac{n_0 \varepsilon^2 R}{C}(1 - e^{-2t/RC}), \qquad (2.31)$$

$$\mathbf{E}\{[V'(t)]^m\} = n_0 \sum_{k=0}^{m-1} \left(-\frac{\varepsilon}{C}\right)^{m-k} \cdot$$

$$\mathbf{E}\{[V'(t)]^k\} \frac{1 - e^{-t(m-k)/RC}}{(m - k - 1)!} RC \frac{(m - 1)!}{k!}. \quad (2.32)$$

From the equations above, we notice that the relative variance of the voltage is

$$\mathrm{Var}(V'(t)) = \frac{1}{2n_0 RC} \frac{1 - e^{-2t/RC}}{(1 - e^{-t/RC})^2}$$

$$\simeq \frac{1}{2n_0 RC} \qquad (2.33)$$

for large t. This result shows that, the larger the value of RC (the time constant of the circuit), the smaller is the fluctuation in voltage and that the fluctuations are important when n_0 (and hence the average current) is small. These results were obtained by Moyal [9]. Correlations of $V'(t)$ at different t's can be obtained from (2.27) by taking appropriate functional derivatives with respect to θ.

Another example of the first-order equation is provided by the Langevin's equation

$$\frac{dy}{dt} + \beta y = x(t). \tag{2.34}$$

Equation (2.34) can describe the flow of current in a simple linear network, the applied electromotive force being the thermal noise voltage $x(t)$ arising from the resistance. It can also describe the motion of a Brownian particle. To describe such situations, $x(t)$ should possess certain special properties. This can be achieved in a number of ways. The usual description is in terms of a Gaussian random process. There is another mode of description, due to Ramakrishnan [10], in terms of a basic random process where the transition probability $R(x|x')$ defined by (2.4) of Chapter 1 is assumed to be a function of x only. In such a case,

$$\int_x R(x|x')\, dx = \text{a constant} = a, \tag{2.35}$$

and the total number of transitions undergone by the random variable $x(t)$ in any interval $(0, t)$ obeys a Poisson distribution with parameter a. Thus the states $x(t)$ before and after a transition are completely uncorrelated and, if a is assumed to very large, $x(t)$ is a wildly fluctuating quantity. If $x(t)$ takes a value x_0 at t_0, the probability frequency function of $x(t)$ at any $t > t_0$ is independent of x_0 if $t - t_0 \gg 1/a$. Other features of this model are discussed at great length in Chapter 4, where we deal with Brownian motion. The relevant point we wish to project is that y as described by (2.34) can be expressed as a linear functional of x, and as such the moments and correlations of y can be expressed as weighted integrals of the correlations of x. For instance, the correlation of y of degree n is given by

$$\mathbf{E}\{y(t_1)y(t_2)\ldots y(t_n)\} = \int_0^{t_1} \int_0^{t_2} \cdots \int_0^{t_n} e^{-\beta[(t_1-\tau_1)+(t_2-\tau_2)+\ldots+(t_n-\tau_n)]}\, .$$

$$\mathbf{E}\{x(t_1)x(t_2)\ldots x(t_n)\}\, d\tau_1\, d\tau_2 \ldots d\tau_n. \tag{2.36}$$

As an example of a second-order differential equation, we can consider an

LCR circuit driven by a thermal noise voltage. The differential equation governing the process is given by

$$\frac{d^2y}{dt^2} + \beta\frac{dy}{dt} + w_0{}^2y = x(t). \tag{2.37}$$

On the basis of the special basic random process characterized by (2.35), Eq. (2.37) can be studied and explicit expressions for the moments and correlations of y can be obtained.

An example of the third-order differential equation is provided by the oscillations of a suspended coil galvanometer (see, for example, Jones and McCombie [11]). If the torsional constant is C and the flux linkage G, the galvanometer deflection θ (which is assumed proportional to the current) satisfies the equations

$$I\frac{d^2\theta}{dt^2} + K\frac{d\theta}{dt} + C\theta = F(t) + Gi,$$

$$L\frac{di}{dt} + Ri = E(t) - G\frac{d\theta}{dt}, \tag{2.38}$$

where $F(t)$ is the random couple and is independent of $E(t)$, the electromotive force in the coil. The system of equations (2.38) can be put into the form (2.19) and the solution can be written in the form (2.21). In this case it is reasonable to assume that $E(t)$ and $F(t)$ are wildly fluctuating and hence can be treated in exactly the same manner as (2.34).

Equations similar to (2.38) can also describe the Brownian motion of electrometers arising from thermal voltage fluctuation and random mechanical torque. On the basis of a Gaussian model, this problem was dealt with by Milatz and Van Zolinger [12] who have obtained the mean square values of the deflection. If we represent the thermal voltage and mechanical torque by appropriate fluctuating fields of the type described by (2.35), higher moments and correlations can be obtained by the use of equations similar to (2.36). Examples of higher-order differential equations arise in the theory of automatic control systems (see, for example, Sundstrom [13]), and these systems can be discussed from the standpoint of Section 2.2.

2.4. Fokker–Planck Equations

In Section 2.3 we encountered a number of physical processes described by ordinary differential equations containing a random forcing term. In many of these situations the random forcing term, which we may denote $x(t)$, possesses the following characteristics:

(i) The average value of $x(t)$ is zero.

(ii) The values of $x(t)$ at two different times, t_1 and t_2, are not correlated at all except for small values of $|t_1 - t_2|$. In other words,

$$E\{x(t_1)x(t_2)\} = \phi(|t_1 - t_2|), \tag{2.39}$$

where $\phi(x)$ is a function with a very sharp maximum at $x = 0$, $\phi(x)$ being very small for $x \neq 0$.

(iii) All correlations of $x(t)$ of odd order vanish:

$$E\{x(t_1)x(t_2) \ldots x(t_{2n+1})\} = 0. \tag{2.40}$$

(iv) All correlations of $x(t)$ of even order can be written as sums of products of second-order correlation:

$$E\{x(t_1)x(t_2) \ldots x(t_{2n})\} = \sum_{\text{all pairs}} E\{x(t_i)x(t_j)\}E\{x(t_k)x(t_l)\} \ldots.$$

Such a process is known as the Uhlenbeck–Ornstein [14] process and has been extensively studied in connection with the theory of Brownian motion (see, for example, Chandrasekhar [15]). We wish to show that, if $y(t)$ satisfies a differential equation of the type (2.19) where $A(t)$ is a constant matrix, it is possible to obtain a partial differential equation for the joint probability frequency function of $y(t)$.

Let us consider the first-order Langevin equation

$$\frac{dy}{dt} + \beta y = x(t), \tag{2.34}$$

where $x(t)$ is assumed to represent the Uhlenbeck–Ornstein process. Thus $y(t)$ can be written in the form

$$y(t) = y_0 e^{-\beta t} + \int_0^t e^{-\beta(t-t')} x(t') \, dt', \tag{2.42}$$

where $y(t)$ takes the value y_0 with probability 1. The first two moments of $y(t + \Delta) - y(t)$ are given by

$$E\{y(t + \Delta) - y(t)\} = (e^{-\beta\Delta} - 1)E\{y(t)\}$$
$$= -\beta \, \Delta E\{y(t)\} + o(\Delta), \tag{2.43}$$

$$E\{[y(t + \Delta) - y(t)]^2\}$$
$$= (e^{-\beta\Delta} - 1)^2 E\{[y(t)]^2\} +$$
$$2D \, \Delta + 2(e^{-\beta\Delta} - 1)E\left\{y(t) \int_t^{t+\Delta} e^{-\beta(t+\Delta-t')} x(t') \, dt'\right\}$$
$$= 2D \, \Delta + o(\Delta), \tag{2.44}$$

where we have approximated $\phi(t)$ defined by (2.39) by

$$\phi(t) = 2D \, \delta(t). \tag{2.45}$$

4

It should be noted that, in obtaining the second moment of $y(t + \Delta) - y(t)$, we have not *assumed* that the third term on the right-hand side of (2.44) vanishes, as is generally believed (see, for example, Gray and Caughey [16]). On the other hand, direct substitution of the integral form of $y(t)$ leads to

$$\mathbf{E}\left\{ y(t) \int_t^{t+\Delta} e^{-\beta(t+\Delta-t')} x(t')\, dt' \right\} = 0, \tag{2.46}$$

a result which is a consequence of the extreme short-range nature of the second-order correlation of $x(t)$. This point is discussed further in Chapter 4.

Next we observe that all higher moments of $y(t + \Delta) - y(t)$ vanish faster than Δ. This fact enables us to obtain a simple partial differential equation for $\pi(y|y_0; t)$, the probability frequency function of y. If we assume that the process $y(t)$ is Markovian* and homogeneous with respect to t, we have the Chapman–Kolmogorov equation

$$\pi(y|y_0; t + \Delta) = \int dz \pi(z|y_0; t) \pi(y|z; \Delta). \tag{2.47}$$

Next we consider the integral

$$\int dy P(y) \frac{\partial \pi(y|y_0; t)}{\partial t} = \lim \frac{1}{\Delta} \, P(y)[\pi(y|y_0; t + \Delta) - \pi(y \mid y_0; t)]\, dy \tag{2.48}$$

where $P(y)$ is an arbitrary smooth function of y going to zero as $u \to \pm \infty$. Substituting the Kolmogorov equation in the right-hand side of (2.48) and expanding $\pi(z|y_0; t)$ about y, we obtain

$$\frac{\partial \pi(y|y_0; t)}{\partial t} = - \frac{\partial}{\partial u}(\pi(y|y_0; t)\beta y) + D\frac{\partial^2}{\partial y^2}\, \pi(y|y_0; t), \tag{2.49}$$

an equation generally known as the Fokker–Planck equation (see [14]). The solution of (2.49) can be identified with a Gaussian distribution with $y_0 e^{-\beta t}$ as the mean and $D(1 - e^{-2\beta t})$ as the mean square deviation.

An exactly similar treatment is possible for the general linear equation of the form (2.19). Wang and Uhlenbeck [17] have dealt with the system of linear equations with constant coefficients, when the different components of $x(t)$ are statistically independent, each of them constituting a Gaussian random process (or, equivalently, an Ornstein–Uhlenbeck process). They have shown that in the general case $y_i(t)$, $i = 1, 2, \ldots, n$, forms a multidimensional Gaussian random process. A more detailed account can be found in the monograph of Middleton [18] which gives a comprehensive account of the statistical properties of communication systems excited by a Gaussian random field.

* The Markovian property at this stage is still an assumption. The justification is provided in Chapter 4.

3. GENERAL STOCHASTIC LINEAR DIFFERENTIAL EQUATIONS

So far we have considered linear differential equations with random forcing terms. However, there arise a number of situations in Physics and Engineering wherein we encounter linear differential equations with stochastic coefficients. The one-dimensional motion of a Brownian particle in a medium with time varying viscosity is an example of such a situation. Another example is provided by the lateral motion of an elastic rod subject to random axial loading. Since the general problem of differential equations with stochastic coefficients is by itself fairly complicated, and not much headway has been made in the direction of the solution of such equations, we have ventured to deal with some particular problems that can be solved in a satisfactory manner. Linear stochastic systems have been studied since about 1960 mainly by specialists in vibration who have made extensive use of Fokker–Planck equations. In fact, there is a general feeling among vibration theorists (see, for example, Caughey and Dienes [19], Ariaratnam and Graefe [20]) that, if each of the coefficients represents a Gaussian random white noise, the Fokker–Planck equation can be set up for the relevant probability frequency function. While the approach through the Fokker–Planck equation can be rigorously justified for a first-order linear differential equation, it is indeed difficult to extend this argument for a general higher-order linear stochastic system (see, for example, Liebowitz [21]). In Sections 3.1 and 3.2 we outline the method of obtaining the Fokker–Planck equation in the case of a first-order linear equation and project the difficulty in treating a general linear system. In Section 3.3 we consider as a simple example of the first-order differential equation a model for emulsion polymerization and show how the relevant p.f.f. can be obtained.

3.1. First-Order Linear Equation

Let us consider the differential equation

$$\frac{dy}{dt} + a(t)y = x(t), \tag{3.1}$$

where $a(t)$ and $x(t)$ are random variables. If $a(t)$ and $x(t)$ represent two independent basic random processes of the type defined in Section 2.1 of Chapter 2, it is easy to arrive at a partial differential equation of the Kolmogorov type for the joint p.f.f. of $y(t)$, $a(t)$, and $x(t)$. Further progress can be made only if the basic random processes possess simpler structure. A large class of physical processes can be represented by (3.1) when $a(t)$ and $x(t)$ are Gaussian white noise. Without loss of generality, let us assume that the mean values of $x(t)$ and $a(t)$ are zero. Then y can be written in the form

$$y(t) = y_0 e^{-b(t)} + \int_0^t x(t') e^{-b(t)+b(t')} \, dt', \tag{3.2}$$

$$b(t) = \int_0^t a(t') \, dt', \tag{3.3}$$

where $y(t)$ takes the value y_0 at $t = 0$. In view of the Gaussian nature of $x(t)$ and $a(t)$, it is worthwhile to investigate whether the p.f.f. of $y(t)$ satisfies a Fokker–Planck equation. With this objective we evaluate the first few moments of $y(t + \Delta) - y(t)$ conditional on $y(t) = y$.

The conditional mean value of $y(t + \Delta) - y(t)$ is given by

$$E\{y(t + \Delta) - y(t)|y\} = E\left\{y\left[\exp\left(-\int_t^{t+\Delta} a(t') \, dt'\right) - 1\right]\right\}$$

$$= yD_{11}\Delta + o(\Delta), \tag{3.4}$$

where D_{11} is connected to the second-order correlation of $a(t)$ by

$$E\{a(t_1)a(t_2)\} = 2D_{11}\,\delta(t_2 - t_1). \tag{3.5}$$

The conditional mean square value of $y(t + \Delta) - y(t)$ is given by

$$E\{[y(t + \Delta) - y(t)]^2|y\} = (D_{11}y^2 - 2D_{10}y + D_{00})\Delta + o(\Delta), \tag{3.6}$$

where D_{10} and D_{00} are defined by

$$E\{a(t_1)x(t_2)\} = 2D_{10}\,\delta(t_2 - t_1),$$
$$E\{x(t_1)x(t_2)\} = 2D_{00}\,\delta(t_2 - t_1). \tag{3.7}$$

If $x(t)$ and $a(t)$ are statistically independent, then $D_{10} = 0$. It can be easily proved that higher-order conditional moments vanish faster than Δ. Thus we can use the method outlined in Section 2.4 to conclude that $\pi(y|y_0; t)$, the p.f.f. of $y(t)$, satisfies the Fokker–Planck equation

$$\frac{\partial \pi(y|y_0;t)}{\partial t} = -\frac{\partial}{\partial y}(D_{11}y\pi) + \frac{\partial^2}{\partial y^2}(D_{11}y^2 - 2D_{10}y + D_{00})\pi. \tag{3.8}$$

Equation (3.8) differs from the usual (incorrect) equation obtained by vibration theorists (see, for example, Bogdanoff and Kozin [22], Ariaratnam and Graefe [20]), the difference arising from the presence of the extra term $(\partial/\partial y)(D_{11}y\pi)$. Thus, many of the results obtained by vibration and structure theorists are open to question. On the other hand, results obtained on the basis of (3.8) will truly represent the properties of the linear system described by (3.1).

3.2. Higher-Order Linear Systems

A general stochastic linear system can be described by

$$\frac{d}{dt}\mathbf{y} + A(t)\mathbf{y} = \mathbf{x}(t), \tag{3.9}$$

where y is a vector with n components, $y_1, y_2, \ldots, y_n$, and $A(t)$ is a $n \times n$ random matrix with elements $a_{ij}(t)$. In general, $x(t)$ can be taken to be a random vector with $x_1(t), x_2(t), \ldots, x_n(t)$. If $x_i(t)$ and $a_{ij}(t)$ represent differentiable random variables by an obvious extension of some of the results of Doob [23] and Ito [24] relating to stochastic integrals, it is possible to arrive at a Fokker–Planck equation for the joint p.f.f. of y_i. However, if the random variables $a_{ij}(t)$ and $x_i(t)$ represent Gaussian white noise characterized by delta function correlations, it is not possible to arrive at a Fokker–Planck equation because of the self-coupling of the y_i's. If the self-coupling of the variables y_i is ignored, we may obtain a Fokker–Planck equation. This has been done by Liebowitz [21] and Ariaratnam and Graefe [20], and their results are open to serious doubt. We elucidate this point by considering a simple second-order equation of the form

$$\frac{d^2y}{dt^2} + a_1\frac{dy}{dt} + a_2(t)y = x(t), \tag{3.10}$$

where a_1 is a deterministic coefficient and $a_2(t)$ and $x(t)$ represent Gaussian white noise. The differential equation can be written as a system:

$$y_1 = \dot{y},$$
$$\frac{dy_1}{dt} + a_1 y_1 + a_2(t)y = x(t), \tag{3.11}$$

which can be written in the integral form

$$y = y(0) + \int_0^t y_1(t')\, dt',$$
$$y_1 = y_1(0)e^{-a_1 t} + \int_0^t [x(t) - a_2(t')y(t')]e^{-a_1(t-t')}\, dt', \tag{3.12}$$

where $y(0)$ and $y_1(0)$ are the initial values of y and dy/dt which may be assumed to be fixed. The set of integral equations (3.12) show that it is not possible to calculate, as in Section 3.1, the conditional moments of $y(t + \Delta) - y(t)$ in view of the coupling of $a(t)$ and $y(t)$ over the interval $(t, t + \Delta)$.

Thus it appears that the stochastic linear equation in its general form is a challenging one, and not much headway can be made except under very simplifying assumptions. One such case has been studied for the equation of the form

$$L_\alpha y \equiv a_n(t)\frac{d^n y}{dt^n} + a_{n-1}\frac{d^{n-1}y}{dt^{n-1}} + \ldots + a_0 y(t) = P(t), \tag{3.13}$$

where the random variable $P(t)$ has the form Mx (M is a differential operator of the type L_α). By writing

$$a_i(t) = \mathbf{E}\{a_i(t)\} + \alpha_i(t), \qquad i = 0, 1, 2, \ldots, n, \tag{3.14}$$

Samuels and Eringen [25] converted (3.13) into the form

$$Ly = P(t) - L_\alpha y, \tag{3.15}$$

where L is the operator obtained from L_α by replacing each $a_i(t)$ by $E\{a_i(t)\}$. A perturbation method is then developed by assuming

$$y(t) = y_0(t) + \varepsilon y_1(t) + \varepsilon^2 y_2(t) + \ldots,$$
$$\alpha_i(t) = \varepsilon \tilde{\alpha}_i(t), \qquad i = 0, 1, 2, \ldots, n, \tag{3.16}$$

and the system (3.15) is replaced by the set of equations

$$Lu_0(t) = P(t),$$
$$Lu_i(t) = - P_{\alpha, i-1}(t),$$
$$P_{\alpha, i-1}(t) = L_\alpha u_{i-1}(t), \tag{3.17}$$

which leads to

$$u_0(t) = \int_0^t g(t, s) P(s) \, ds + \sum_{k=1}^n c_k w_k(t),$$

$$u_i(t) = \int_0^t g(t, s) P_{\alpha, i-1}(s) \, ds, \tag{3.18}$$

where $g(t, s)$ is the Green's function associated with the operator L and the w_k's form a fundamental set of independent solutions of the deterministic equation $Ly = 0$. Equations (3.18) provide a basis for the discussion of the statistical properties of $y(t)$. The perturbation method will be successful only if the stochastic coefficients are small or are slowly varying. The success of this method has been demonstrated for one or two specific instances by Samuels and Eringen.

3.3. A Simple Example

Let us consider a stochastic model for emulsion polymerization proposed by Stanley Katz [26]. We shall assume that monomer diffuses from an ambient liquor into blobs of monomer-polymer mixture and polymerizes at a rate depending on its own concentration. The random element is introduced by assuming that the radicals from the ambient liquor diffuse into blobs in a Poissonian way, their successive arrivals alternately triggering and inhibiting the polymerization. The equation governing the process can be represented by

$$\frac{dc}{dt} = h(c_0 - c) - kc f(t), \tag{3.19}$$

where c is the monomer concentration at any time and c_0 is the fixed monomer concentration in the ambient liquor, h being the mass transfer coefficient governing the diffusion of monomer into the blob. The rate of polymerization is taken proportional to the monomer concentration in the blob with the proportionality factor k. The stochastic element in the differential equation is $f(t)$, which takes the values 0 and 1 corresponding to the starting and stopping of the polymerization at random times of arrivals of radicals from the ambient so that

$$f(t) = \tfrac{1}{2} - \tfrac{1}{2}(-)^{N(t)}, \tag{3.20}$$

where $N(t)$ is the random variable representing the Poisson process. The variable of interest from the viewpoint of polymer chemistry is $r(t)$, the rate of polymerization, and is given by

$$r(t) = kcf(t). \tag{3.21}$$

The p.f.f. of c was studied by Srinivasan [27], who also obtained the moments of the random variable $r(t)$. We shall briefly outline the method of solution.

Let $\pi_{0(1)}(c, t)\, dc$ denote the joint probability that the random variable $f(t)$ takes the value 1(0) and $c(t)$ has a value between c and $c + dc$. We find

$$\frac{\partial \pi_0(c, t)}{\partial t} = -(\lambda - h)\pi_0(c, t) + \lambda \pi_1(c, t) - h(c_0 - c)\frac{\partial \pi_0(c, t)}{\partial c}, \tag{3.22}$$

$$\frac{\partial \pi_1(c, t)}{\partial t} = -(\lambda - h - k)\pi_1(c, t) + \lambda \pi_0(c, t) - \{h(c_0 - c) - kc\}\frac{\partial \pi_1(c, t)}{\partial c} \tag{3.23}$$

with the initial conditions

$$\pi_0(c, 0) = \delta(c - c_0), \qquad \pi_1(c, 0) = 0, \tag{3.24}$$

corresponding to the physical situation that the blob is at the same monomer concentration as the ambient initially. If $\pi(r, t)$ is the p.f.f. of the rate of polymerization, then

$$\pi(r, t) = \int_c \pi_0(c, t)\, dc\, \delta(r) + \int_c \pi_1(c, t)\, \delta(r - kc)\, dc. \tag{3.25}$$

Defining

$$E\{r^n\} = \int r^n \pi(r, t)\, dr,$$

we obtain

$$E\{r^n\} = \int \pi_1\left(\frac{r}{k}, t\right)\frac{r^n}{k}\, dr. \tag{3.26}$$

Thus the moments of r can be obtained directly from the moments of $\pi_1(r, t)$. Although it is difficult to obtain the explicit solution of (3.22) and (3.23), it is quite easy to obtain the moments of the functions π_0 and π_1. The details of the calculations will be found in [27].

The example shows the inapplicability of general methods even for a first-order differential equation having a simple structure. We have occasion to study similar examples in Chapter 5.

4. NONLINEAR STOCHASTIC EQUATIONS

With the help of the theory of linear equations subject to random excitations as dealt with in Section 2 of this chapter, we can predict the statistical behavior of the response of linear mechanical systems to random excitation. On the other hand, the nonlinear systems cannot be analyzed in terms of the general techniques that have been developed predominantly for linear processes. Although a large number of people—the kinetic theorists, control theorists, and vibration and structure analysts—have contributed to the understanding of this subject, it is very difficult to visualize any general pattern or method of solution. In this section we demonstrate the inapplicability of the Fokker–Planck approach and present the quasilinearization technique developed by Caughey and his coworkers during the last few years. In Section 4.2, we illustrate the method by considering a hard spring subject to random excitation.

To be specific, let us consider the first-order nonlinear equation

$$\frac{dy}{dt} + ay + by^2 = x(t), \tag{4.1}$$

where a and b are constants and $x(t)$ represents the random force to which the system is subjected. It is generally believed that the p.f.f. of $y(t)$ satisfies a Fokker-Planck equation if $x(t)$ is a Gaussian white noise process (see, for example, Lyon [28], Caughey [29]). To see why such a viewpoint is open to serious doubt, we rewrite (4.1) in the integral form

$$y = y_0 \exp\left[- at - b \int_0^t y(t')\, dt'\right] + \int_0^t x(t') \exp\left[- a(t - t') - b \int_{t'}^t y(t'')\, dt''\right] dt'. \tag{4.2}$$

The first step consists of the evaluation of the conditional moments of $y(t + \Delta) - y(t)$ conditional on $y(t) = y$. The random variable $y(t + \Delta) - y(t)$ can be written

$$y(t + \Delta) - y(t) = y(t)\left\{\exp\left[- a\Delta - b \int_t^{t+\Delta} y(t')\, dt'\right] - 1\right\} + \int_t^{t+\Delta} x(t')\left\{\exp\left[- a(t + \Delta - t') - b \int_{t'}^{t+\Delta} y(t'')\, dt''\right]\right\} dt', \tag{4.3}$$

where the nonlinear nature of the original equation is reflected by the presence of the values of y in the range $(t, t + \Delta)$. In the usual derivation of a Fokker–Planck equation, the first two moments of $y(t + \Delta) - y(t)$ are estimated as

$$E\{y(t + \Delta) - y(t)|y\} = y[e^{-a\Delta - yb\Delta} - 1] + o(\Delta)$$
$$= - (ay + by^2) \Delta + o(\Delta),$$
$$E\{[y(t + \Delta) - y(t)]^2|y\} = D_{00} \Delta + o(\Delta), \tag{4.4}$$

where D_{00} is defined by (3.7). Equations (4.4) lead to a Gaussian distribution for $y(t)$. However, this procedure is not a sound one because the internal correlation that usually arises from the nonlinear nature, and is expressed as an integral of y under the exponent in the range $(t, t + \Delta)$ in (4.3), is completely neglected. In fact, in the process of evaluating the moments of $y(t + \Delta) - y(t)$ as expressed by (4.4), the Gaussian (nonwhite) nature is assumed. It is our contention that the Gaussian (nonwhite) nature of the process $y(t)$ can be proved only if $y(t)$ satisfies a linear equation. On the other hand, for nonlinear equations, other techniques similar to quasilinearization should be used to arrive at the statistical properties of $y(t)$.

4.1. Equivalent Linearization Technique

The purpose of this section is to discuss an approximate method of dealing with nonlinear differential equations. This method is due to Caughey [30] and his collaborators who systematically analysed nonlinear phenomena in solid mechanics by considering corresponding equivalent linear phenomena. Such a technique has been originally adopted for deterministic problems by Kryloff and Bogoliuboff [31]. To illustrate the development of the theory, let us consider a nonlinear oscillator subjected to Gaussian random excitation not necessarily of the white noise type described by

$$\ddot{y} + \beta\dot{y} + \omega_0{}^2 y + \eta g(y, \dot{y}, t) = x(t), \tag{4.5}$$

where β and η are small and thus the system can be considered to be weakly nonlinear and lightly damped. The technique consists in replacing (4.5) by

$$\ddot{y} + \beta_{eq}\dot{y} + \omega_{eq}^2 y + e(y, \dot{y}, t) = x(t), \tag{4.6}$$

where β_{eq} is the equivalent "linear damping" coefficient per unit mass and ω_{eq}^2 is the "equivalent linear stiffness" coefficient. $e(y, \dot{y}, t)$ is the error or equation deficiency term and, if it is neglected, (4.6) is linear and can be readily solved. The smaller the magnitude of the error term, the smaller is the error in neglecting it. Naturally, the choice of ω_{eq} and β_{eq} is the set of values that make $e(y, \dot{y}, t)$ a minimum. Thus we can obtain the values of ω_{eq} and β_{eq} by minimizing the mean squared error. $e(y, \dot{y}, t)$ is explicitly given by

$$e(y, \dot{y}, t) = (\beta - \beta_{eq})\dot{y} + (\omega_0 - \omega_{eq}^2)y + \eta g(y, \dot{y}, t), \tag{4.7}$$

so that the mean squared error can be written

$$\overline{e^2} = \lim_{T \to \infty} \frac{1}{2T} \int_{-T}^{T} [(\beta - \beta_{\text{eq}})\dot{y} + (\omega_0{}^2 - \omega_{\text{eq}}^2)y + \eta g(y, \dot{y}, t)] \, dt. \quad (4.8)$$

The minimization conditions yield

$$\beta_{\text{eq}} = \beta + \frac{\overline{\eta y g(y, \dot{y}, t)}}{\overline{\dot{y}^2}},$$

$$\omega_{\text{eq}}^2 = \omega_0{}^2 + \frac{\overline{\eta y g(y, \dot{y}, t)}}{\overline{\dot{y}^2}}. \quad (4.9)$$

It can be shown that the conditions (4.9) do indeed define a minimum for $\overline{e^2}$. If we assume that the process is ergodic, then the time averages can be replaced by the ensemble averages. A self-consistent method of estimating the parameters β_{eq} and ω_{eq}^2 consists in using the linear form of (4.6) as the basis for the evaluation of the ensemble averages, yielding two simultaneous equations for β_{eq} and ω_{eq}^2.

This technique can be extended to systems having higher degrees of freedom. Caughey has dealt with such systems, the nonlinearity being confined to the displacements alone.

4.2. Hard Spring Oscillator

The random motion of a hard spring oscillator driven by Gaussian white noise can be studied by the equivalent linearization technique. The problem was studied by Lyon [32], who compared the solution with that obtained via a Fokker–Planck equation. The equation of motion in this case is given by

$$\ddot{y} + 2\alpha\dot{y} + \omega_0{}^2(y + y^3) = x(t), \quad (4.10)$$

where x is a Gaussian white noise with spectral density

$$D = 4\alpha\omega_0{}^2\sigma_D{}^2, \quad (4.11)$$

$\sigma_D{}^2$ being the variance of η with the cubic term absent. To apply equivalent linearization, we rewrite (4.10) in the form

$$\ddot{y} + 2\alpha\dot{y} + \Omega_0{}^2 y + \varepsilon = x(t). \quad (4.12)$$

If we can neglect ε, then y is normal with variance

$$E\{y^2\} = \sigma_d{}^2 \frac{\omega_0{}^2}{\Omega_0{}^2}. \quad (4.13)$$

We assume that ε may be neglected if $E\{\varepsilon^2\}$ is minimized by the proper choice of Ω_0. Adopting the procedure explained in Section 4.1, we find the choice for Ω_0 to be

$$\frac{\Omega_0^2}{\omega_0^2} = 1 + 3E\{y^2\}. \tag{4.14}$$

Substituting (4.14) in (4.13), we can solve for Ω_0^2 or, better, $E\{y^2\}$ directly:

$$E\{y^2\} = -\tfrac{1}{6} + \frac{(1 + 12\sigma_d^2)^{\frac{1}{2}}}{6}. \tag{4.15}$$

Since y is normal, the higher moments of y can be calculated readily. Lyon verified that the exact calculation of the second moment via a Fokker–Planck equation and the approximate moment given by (4.15) do not differ by more than 10 percent.

In the light of the remarks made in the beginning of this section, the agreement of the result within 10 percent is essentially due to the weak nonlinear nature of the phenomenon. If such a result can be obtained for a fairly strong nonlinearity, it would provide a justification for the approach via the Fokker–Planck equation, since in the letter approach the Gaussian nature is already assumed in the course of the derivation of the Fokker–Planck equation.

5. DIFFERENTIAL EQUATIONS WITH RANDOM INITIAL CONDITIONS

Differential equations characterized by random initial or boundary conditions are the easiest to handle from the viewpoint of the theory of stochastic processes. The probabilistic arguments are the simplest because the stochastic properties of the solutions are directly related to the initial conditions. Let us consider the equation

$$Ly(t) = x(t), \tag{5.1}$$

where L is a linear differential operator and $x(t)$ is a deterministic function of t. Suppose that the initial conditions are given by

$$y^{(i)}(0) = \xi_i, \qquad i = 1, 2, \ldots, n, \tag{5.2}$$

where $y^{(i)}(t) = d^i y/dt^i$. If the statistical properties of ξ_i are given, the statistical properties of $y(t)$ can be obtained. We can write the solution of y in the form

$$y(t) = f(t, \xi_1, \xi_2, \ldots, \xi_n). \tag{5.3}$$

If $\pi(\xi_1, \xi_2, \ldots, \xi_n)$ is the joint p.f.f. of the set $[\xi_i]$, the joint p.f.f. of y and its first $(n - 1)$ derivatives can be readily obtained. Denoting the joint p.f.f. by $\rho(y, y_1, y_2, \ldots, y_{n-1}, t)$ from elementary probabilistic arguments, it follows that

$$\rho(y, y_1, y_2, \ldots, y_{n-1}, t) = \pi(\xi_1, \xi_2, \ldots, \xi_n)\frac{\partial(\xi_1, \xi_2, \ldots, \xi_n)}{\partial(y, y_1, y_2, \ldots, y_{n-1})}. \tag{5.4}$$

To obtain ρ explicitly we can solve for $\xi_1, \xi_2, \ldots, \xi_n$ in terms of $y, y_1, y_2,$ $\ldots, y_{n-1}$, using the relation

$$y_i = \frac{d^i}{dt^i} f(t, \xi_1, \xi_2, \ldots, \xi_n), \tag{5.5}$$

and then substitute the solutions in the right-hand side of (5.4). Such a procedure is always possible, at least in principle. An excellent example of a system of differential equations with random initial conditions occurs naturally in the formulation of Statistical Mechanics. Of course, the number of degrees of freedom of such a system is of the order of 10^{23}, and we are normally interested in certain projections of the trajectories of the system and not the trajectories which permit solutions of the form (5.3). In Continuum Mechanics we encounter a class of problems which involve solution of partial differential equations with random boundary conditions, and they are discussed in Chapter 6. Some simple problems involving ordinary differential equations will be found in the monograph of Saaty [33].

REFERENCES

1. A. T. Bharucha-Reid, *Proc. Symp. Appl. Math.*, **16**(1964), 40.
2. A. T. Bharucha-Reid, *Proc. Symp. Theoret. Phys.*, **2**(1966).
3. P. M. Mathews and S. K. Srinivasan, *Proc. Indian Acad. Sci.*, A43(1956), 4.
4. E. T. Whittaker and G. N. Watson, *A Course of Modern Analysis*, Cambridge, 1927,
5. C. A. Coddington and N. L. Levinson, *Theory of Ordinary Differential Equations*. McGraw-Hill, New York, 1955.
6. S. K. Srinivasan, *Z. Angew. Math. Mech.*, **43**(1963), 259.
7. L. Schleisinger, *Handbuch der Theorie der linearen Differentialgleichungen*, Vol. 1, Leipzig, 1895, p. 52.
8. G. Polya, *Trans. Amer. Math. Soc.*, **24**(1922), 312.
9. J. E. Moyal, *J. Roy. Statist. Soc.*, B11(1949), 150.
10. Alladi Ramakrishnan, *Astrophys. J.*, **119**(1954), 443.
11. R. V. Jones and C. W. McCombie, *Philos. Trans.*, A244(1951–52), 205.
12. J. M. W. Milatz and J. J. Van Zolingen, *Physica*, **19**(1953), 181.
13. M. Sundstrom, *Arkiv. Math.*, **2**(1945), 323.
14. G. E. Uhlenbeck and L. S. Ornstein, *Phys. Rev.*, **36**(1930), 823.
15. S. Chandrasekhar, *Rev. Mod. Phys.*, **15**(1943), 1.
16. A. H. Gray and T. K. Caughey, *J. Math. Phys.*, **44**(1965), 288.
17. M. C. Wang and G. E. Uhlenbeck, *Rev. Mod. Phys.*, **17**(1945), 323.
18. D. Middleton, *An Introduction to Statistical Communication Theory*, McGraw-Hill, New York, 1960.
19. T. K. Caughey and J. K. Dienes, *J. Math. Phys.*, **41**(1962), 300.
20. S. T. Ariaratnam and P. W. U. Graefe, *Int. J. Control*, **1**(1965), 239.
21. M. A. Liebowitz, *J. Math. Phys.*, **4**(1963), 852.
22. J. L. Bogdanoff and F. Kozin, *J. Acoust. Soc. Amer.*, **34**(1962), 1063.

23. J. L. Doob, *Ann. Math.*, **43**(1942), 351.
24. K. Ito, On Stochastic Differential Equations, *Mem. Amer. Math. Soc.*, No. 4 (1951).
25. J. C. Samuels and C. A. Eringen, *J. Math. Phys.*, **38**(1959), 83.
26. S. Katz, *J. Soc. Indust. Appl. Math.*, **8**(1960), 368.
27. S. K. Srinivasan, *J. Soc. Indust. Appl. Math.*, **11**(1963), 355.
28. R. H. Lyon, *J. Acoust. Soc. Amer.*, **32**(1960), 716.
29. T. K. Caughey, *J. Acoust. Soc. Amer.*, **35**(1963), 1683.
30. T. K. Caughey, *J. Acoust. Soc. Amer.*, **35**(1963), 1706.
31. N. Kryloff and N. Bogoliuboff, *Introduction to Non-linear Mechanics*, translated by S. Lefschetz, *Ann. of Math. Studies*, No. 11, Princeton University Press, Princeton, N.J., 1947.
32. R. H. Lyon, *J. Acoust. Soc. Amer.*, **32**(1966), 1161.
33. T. L. Saaty, *Modern Non-linear Equations*, McGraw-Hill, New York, 1967.

Chapter 4

BROWNIAN MOTION

1. INTRODUCTION

The Brownian movement was first physically observed by a botanist called
R. Brown [1] who noticed that, when pollen is dispersed in water, the
suspended particles execute a random walk in three dimensions. This
phenomenon was studied by Einstein, who gave an elegant theory in a series
of papers [2] describing the motion of the suspended particles under the
influence of a fluctuating force. A detailed study of this phenomenon involves
the solution of a differential equation with a stochastic term responsible for
the random motion of these particles. The equation proposed by Einstein
leads directly to an analysis of the Kolmogorov type of equations, which
under certain restrictions can be described by the famous Fokker–Planck
equations governing the distribution function of the displacement or velocity
of the Brownian particle. The physical applications of this are legion, covering
a wide range of fields: kinetic theories occurring in various physical situations, plasma physics, astrophysics, quantum noise, motion of lattice chains
with impurities, genetics, and even behavior of structures under stochastic
loads. On the pure mathematical side, this has led to many important
concepts like the Wiener integrals (see, for example, [3]).

The plan of this chapter is as follows. We describe first the diffusion
equation and its discrete version in terms of random walk problem. In
Section 3 we present an account of the general Kolmogorov equations
governing the p.f.f. of the position of the Brownian particle. Since the main
starting point for the theory of Brownian motion is the Langevin equation,
we devote Section 4 to a discussion of the various physical consequences of
the equation. We then discuss at length the basic assumptions and their
implications for Markovian nature and fluctuation dissipation theorem.
Section 6 is devoted to the derivation of the Schrodinger equation from the
viewpoint of Brownian motion.

2. DIFFUSION EQUATION

The diffusion equation, first proposed by Einstein [2], describes the
probability frequency function of the position of the Brownian particle.

This led to the determination of Avogadro's number since the diffusion coefficient occurring in the equation is a function of the temperature and viscosity of the medium as well as the dimensions of the particle. By calculating the mean square displacements of the Brownian particle, we can determine the diffusion coefficient. In this section we outline the main line of reasoning.

We observe first that the Brownian particle suffers 10^{21} collisions per second and that, if it starts from an initial value of its position $x = 0$ at $t = 0$ and reaches the position x at time t, t being large compared to collision times, the cumulative effect of all the impulses, by the central limit theorem, leads to a normal distribution given by

$$f(x, t) = \tfrac{1}{2}(\pi D t)^{-\frac{1}{2}} e^{-x^2/4Dt}, \tag{2.1}$$

where $f(x, t)$ satisfies a diffusion equation of the form

$$\frac{\partial f}{\partial t} = D \frac{\partial^2 f}{\partial x^2}. \tag{2.2}$$

We then calculate D from physical considerations. If particles are spherical and are of radius a, the Stokes law gives the force resisting the motion of the particles as $F = 6\pi a \eta v$, where η is the coefficient of viscosity of the medium and v the velocity of the particle in the medium. Taking into account the flux of particles across a unit area at a given point x caused by diffusion, we have

$$F = -\frac{6\pi a \eta D}{\rho} \frac{\partial \rho}{\partial x}. \tag{2.3}$$

The osmotic pressure of the suspended particles at the temperature T gives a net opposing force canceling F, and hence we have

$$\frac{RT}{N} \frac{\partial \rho}{\partial x} = \frac{6\pi a \eta D}{\rho} \frac{\partial \rho}{\partial x}, \tag{2.4}$$

so that

$$D = \frac{RT\rho}{6N\pi\eta a}. \tag{2.5}$$

This led to the determination of Avogadro's number N. More important than that is that this work of Einstein made way for the emergence of statistical mechanics as a theory having experimental consequences, independent of those of thermodynamics, and gave a fillip to the atomic hypothesis, the strength of the resistance to which cannot be appreciated at this distance in time. The interesting applications of this phenomena include the problems of random walk, the Langevin equations, Fokker–Planck equations, the equations of diffusion, the transition to the macroscopically irreversible nature of diffusion from the microscopically reversible nature of molecular fluctuations, and even problems in stellar dynamics. A detailed study of some of these aspects will be found in the classic article by Chandrasekhar [4]. A lucid account of these methods, which are common to many of these problems, particularly motion of a star amidst galaxies and random flights, can be found in the survey article by Ramakrishnan [5].

A very simple and discrete version of a diffusion process is the one-dimensional random walk in which a particle suffers displacements along a line in the form of a series of steps of equal length, each step being taken either in the forward or the backward direction. We ask the question: What is the probability $p(n|m, s\tau)$ that the particle starting from the nth lattice point goes to m, in time $s\tau$ corresponding to s steps, where $s = 1, 2, \ldots$, and τ is the unit time of each step? We define $Q(n|m)$ as the transition probability that the particle goes from the mth point to the nth in one step. From the Markov nature of the problem, and making repeated applications of the Smoluchowsky equation, we obtain

$$p(n|m, s) = \sum_k p(n|k, s - 1)Q(k, m). \tag{2.6}$$

We can solve this problem for small values of s. For large values of s, we write (2.6) as

$$p(m, s) - p(m, s - 1) = - p(m, k - 1)\sum_k{}' Q(m, k) + \sum_k{}' p(k, s - 1)Q(k, m), \tag{2.7}$$

where the prime on the sum indicates that $m = k$ is excluded from the sum. For equal probability for the right and left steps, the transition probability Q is given by

$$Q(k, m) = \tfrac{1}{2}\delta(m, k - 1) + \tfrac{1}{2}\delta(m, k + 1). \tag{2.8}$$

Equation (2.7) is like the Boltzmann equation in the kinetic theory of gases and has to be solved with the initial condition $p(m, 0) = \delta(n, m)$. In the limit, $s\tau = t$ and $m\Delta = x$, Δ being the width of a step, become continuous variables and (2.7) coupled with (2.8) becomes the diffusion equation:

$$\frac{\partial p(x, t)}{\partial t} = D\frac{\partial^2 p(x, t)}{\partial x^2}, \tag{2.9}$$

where $D = \lim (\Delta^2/2\tau)$.

A variation of the problem, in which the transition probabilities for making a step forward or backward are no longer equal, has the form

$$Q(k, m) = \frac{L + k}{2L}\delta(m, k - 1) + \frac{L - k}{2L}\delta(m, k + 1), \tag{2.10}$$

where L is an integer, corresponding to the fact that there is an attractive center. For this example, if we go now to the continuous limit as we did earlier, we arrive at

$$\frac{\partial}{\partial t}p(x, t) = \beta\frac{\partial}{\partial x}(xp(x, t)) + D\frac{\partial^2}{\partial x^2}p(x, t), \tag{2.11}$$

where $\beta = \lim (1/\tau L)$.

The way in which solutions of these equations for the discrete cases go into the solutions of the continuous cases is discussed by Ornstein *et al.* [6, 7] and Chandrasekhar [4].

3. KOLMOGOROV EQUATIONS

The Markov processes dealt with in the preceding section are, of course, idealized abstractions but very convenient to study various physical situations. In fact, these processes without memory are amenable to mathematical methods which give a great insight into actual systems encountered in practice. The vital starting point in describing a continuous Markov process $x(t)$ is the Smoluchowsky equation or Chapman–Kolmogorov equation obeyed by the function $\pi(x|y, t, t_0)$, the probability that x assumes the value between x and $x + dx$ at time t, starting from y at t_0. Assuming homogeneity in time and setting $t_0 = 0$, we observe that

$$\pi(x|y; t + \Delta) = \int_{-\infty}^{\infty} \pi(z|y; t)\pi(x|z; \Delta)\, dz. \tag{3.1}$$

This is the forward equation; the backward equation is given by

$$\pi(x|y; t) = \int_{-\infty}^{\infty} \pi(z|y; \Delta)\pi(x|z; t - \Delta)\, dz. \tag{3.2}$$

In (3.3), we expand $\pi(x|z; t - \Delta)$ about y:

$$\pi(x|y; t) = \int_{-\infty}^{\infty} \pi(z|y; \Delta)\left\{\pi(x|y, t - \Delta) + (z - y)\frac{\partial\pi}{\partial y} + \right.$$
$$\left. \frac{(z - y)^2}{2!}\frac{\partial^2\pi}{\partial y^2} + \ldots\right\} dz. \tag{3.3}$$

If we also assume that the moments of the changes in the random variable $x(t)$ in time interval Δ are such that

$$\lim_{\Delta\to 0}\frac{1}{\Delta}\int(z - y)\pi(z|y; \Delta)\, dz = a(y), \tag{3.4}$$

and

$$\lim_{\Delta\to 0}\frac{1}{\Delta}\int(z - y)^2\pi(z|y; \Delta)\, dz = b(y), \tag{3.4}$$

5

and, for all $n > 2$,

$$\lim_{\Delta \to 0} \frac{1}{\Delta} \int (z - y)^n \pi(z|y; \Delta) \, dz = 0,$$

we obtain the backward equation

$$\frac{\partial \pi}{\partial t} = \tfrac{1}{2} b(y) \frac{\partial^2 \pi}{\partial y^2} + a(y) \frac{\partial \pi}{\partial y}. \tag{3.5}$$

However, the forward equation (3.1) is more interesting and leads to the well-known Fokker–Planck equation description of the various physical situations. There are different ways of dealing with (3.1) to obtain the Fokker–Planck equation. We describe the method, using the characteristic function of the transition probability $\pi(x|z; \Delta)$ defined by

$$\theta(u; z) = \mathbf{E}\{e^{iu(x-z)}\} = \int e^{iu(x-z)} \pi(x|z; \Delta) \, dx. \tag{3.6}$$

Equally well we can use the inverse transform

$$\pi(x|z; \Delta) = \frac{1}{2\pi} \int e^{-iu(x-z)} \theta(u, z) \, du. \tag{3.7}$$

Substituting the above in (3.1), we have

$$\pi(x|y; t + \Delta) = \frac{1}{2\pi} \int \int e^{-iu(x-z)} \theta(u; z) \pi(z|y; t) \, du \, dz. \tag{3.8}$$

Expanding the characteristic function in terms of the moments of the increments in x, during the interval Δ, we obtain

$$\theta(u; z) = 1 + \sum_{s=1}^{\infty} \frac{(iu)^s}{s!} m_s(z), \tag{3.9}$$

where

$$m_s(z) = \mathbf{E}\{(x - z)^s\}. \tag{3.10}$$

Since

$$\frac{1}{2\pi} \int e^{-iu(x-z)} (iu)^s \, du = -\left(\frac{\partial}{\partial x}\right)^s \delta(x - z), \tag{3.11}$$

we have, on substituting (3.9) in (3.8),

$$\pi(x|y; t + \Delta) = \pi(x|y; t) + \sum_{s=1}^{\infty} \frac{1}{s!} \left(-\frac{\partial}{\partial x}\right)^s [m_s(x) \pi(x|y; t)]. \tag{3.12}$$

It is easy to see that, as $\Delta \to 0$, π satisfies the equation

$$\frac{\partial}{\partial t} \pi(x|y, t) = \sum_{s=1}^{\infty} \frac{1}{s!} \left(-\frac{\partial}{\partial x}\right)^s [K_s(x) \pi(x|y; t)] \tag{3.13}$$

with $K_s = \lim_{\Delta \to 0}(m_s(x))/\Delta$. If all the terms on the right-hand side of (3.13) do not vanish, each moment m_s is to be of the form $K_s(x)\Delta$. If the coefficients K_2, $K_4, \ldots$ are equal to zero, (3.13) is of the form

$$\frac{\partial}{\partial t}\pi(x|y; t) = -\frac{\partial}{\partial x}[K_1(x)\pi(x|y; t)] + \tfrac{1}{2}\frac{\partial^2}{\partial x^2}[K_2(x)\pi(x|y; t)] \quad (3.14)$$

with the initial condition $\pi(x|y, 0) = \delta(x - y)$. Introducing the probability current (see Stratonovich [8])

$$j(x) = K_1(x)\pi(x|y; t) - \tfrac{1}{2}\frac{\partial}{\partial x}[K_2(x)\pi(x|y; t)], \quad (3.15)$$

the Fokker–Planck equation expresses the conservation of probability in the form

$$\frac{\partial \pi(x)}{\partial t} + \frac{\partial j(x)}{\partial x} = 0. \quad (3.16)$$

In every particular realization of a Markov process the trajectory $x(t)$ can be thought of as being swept out by a representative point which moves along the x-axis just like a Brownian particle or a particle undergoing diffusion governed by an equation of the type (3.14). The existence and uniqueness of solutions of the Fokker–Planck equations employ the theory of semigroups and are treated by Feller [9], Hille and Phillips [10], and Yosida [11].

There are two well-known diffusion processes which are particular cases of the Fokker–Planck equation. If we assume that $K_2 = \text{constant} = 2D$ and $K_1 = 0$, we obtain the equation of the Wiener–Levy process given by

$$\frac{\partial \pi(x; t)}{\partial t} = D\frac{\partial^2 \pi(x; t)}{\partial x^2}, \quad (3.17)$$

where D is the coefficient of diffusion. In addition to representing Brownian motion, it also represents heat conduction, where $\pi(x; t)$ represents the amount of heat in the system at x at time t. If the coefficients K_1 and K_2 are such that the Fokker–Planck equation (3.14) is transformed as

$$\frac{\partial \pi(x; t)}{\partial t} = \beta\frac{\partial(x\pi(x; t))}{\partial x} + D\frac{\partial^2 \pi(x; t)}{\partial x^2}, \quad (3.18)$$

corresponding to Brownian motion with viscous forces included, we have what is called the Uhlenbeck–Ornstein process leading to a solution of the form

$$\pi(x|y; t) = \frac{1}{(2\pi\sigma^2)^{\frac{1}{2}}} \exp\left[\frac{-(x - \bar{y})^2}{2\sigma^2}\right], \quad (3.19)$$

where

$$\bar{y} = ye^{-\beta t}, \qquad \sigma^2 = \frac{D}{\beta}[1 - \exp(-2\beta t)]. \tag{3.20}$$

The analysis above can be easily extended to a multidimensional process, that is, a Markov process which consists of several random functions $x_1(t), x_2(t), \ldots, x_m(t)$ which are independent of each other. The probability density $\pi(x_1, x_2, \ldots, x_m; t)$ satisfies the multidimensional Fokker–Planck equation

$$\frac{\partial \pi(x_1, x_2, \ldots, x_m; t)}{\partial t} = - \sum_{\alpha=1}^{m} \frac{\partial}{\partial x_\alpha}[K_\alpha(x_1, \ldots, x_m)\pi(x_1, \ldots, x_m; t)] +$$

$$\tfrac{1}{2}\sum \frac{\partial^2}{\partial x_\alpha \partial x_\beta}[K_{\alpha\beta}(x_1, \ldots, x_m)\pi(x_1, \ldots, x_m; t)], \tag{3.21}$$

where the coefficients K_α and $K_{\alpha\beta}$ are defined as

$$K_\alpha = \lim_{\Delta \to 0} \frac{1}{\Delta}\mathbf{E}\{x_\alpha(t + \Delta) - x_\alpha\},$$

$$K_{\alpha\beta} = \lim_{\Delta \to 0} \frac{1}{\Delta}\mathbf{E}\{(x_\alpha(t + \Delta) - x_\alpha)(x_\beta t + \Delta) - x_\beta)\}. \tag{3.22}$$

Though the applications of the forward equation (3.18) to particular situations are discussed in Section 4.5 we indicate here the utility of the backward equation (3.5) in processes which eventually terminate. This category includes first passage time problems. We shall be concerned with one-dimensional problems within the region $x_1 < x < x_2$. Let $x(t)$ be a Markov process; we are interested in finding the time T_f in which the process reaches the boundaries, x_1, x_2, for the first time. We then calculate the mean first passage time $\bar{T}_f$. Let us define $\tilde{\pi}(x|x_0, t, t_0)$ the probability that the process has values between x and $x + dx$ at time t, without ever having touched the boundaries. Then the integral

$$S(t) = \int_{x_1}^{x_2} \tilde{\pi}(x; t)\, dx \tag{3.23}$$

gives the probability that the process has not reached the boundaries during the interval $[t, t_0]$. At the initial time t_0, the process is to lie somewhere within the boundaries to start the process, and so the following normalization should hold

$$S(t_0) = \int_{x_1}^{x_2} \tilde{\pi}(x; t_0)\, dx = 1. \tag{3.24}$$

At subsequent times t, $S(t)$ should be less than unity, and after infinite time the process should have reached the boundaries and $S(\infty)$ should be equal to zero. If the process starts from the point x_0, $\tilde{\pi}(x; t_0) = \delta(x - x_0)$, the mean time spent within the boundary is given by

$$\overline{T}_f = \int_{t_0}^{\infty} S(t, x_0)\, dt, \tag{3.25}$$

where

$$S(t, x_0) = \int_{x_1}^{x_2} \tilde{\pi}(x|x_0, t - t_0)\, dx. \tag{3.26}$$

Let us now integrate the backward equation (3.5) over x in the allowed region to obtain

$$\frac{\partial S(t, x_0)}{\partial t} = a(x_0)\frac{\partial S}{\partial x_0} + \tfrac{1}{2}b(x)\frac{\partial^2 S}{\partial x_0{}^2}. \tag{3.27}$$

Integrating over t, from t_0 to ∞, we obtain the equation for $\overline{T}_f$ straightaway as

$$-1 = a(x_0)\frac{d\overline{T}_f}{dx_0} + \tfrac{1}{2}b(x_0)\frac{\partial^2 \overline{T}_f}{\partial x_0{}^2}. \tag{3.28}$$

Since the mean first passage time is zero if x_0 is either x_1 or x_2 (the boundaries), we have

$$\overline{T}_f(x_1) = \overline{T}_f(x_2) = 0. \tag{3.29}$$

For constant values of $(b(x_0)/2) = R$, we have the solution

$$\overline{T}_f(x_0) = \frac{1}{R}(x_0 - x_1)(x_2 - x_0). \tag{3.30}$$

4. LANGEVIN EQUATION

The Langevin equation is a typical differential equation governing many physical phenomena subject to random forces. A detailed study of the equation was carried out by Ornstein, Wang and Uhlenbeck [6, 7]. In the Sections 4.1 and 4.2 we carry out an analysis of the equation leading to the Fokker–Planck equation and its solution. Section 4.3 contains the derivation of the backward equation.

4.1. Solution of the Langevin Equation

The motion of a Brownian particle suffering random changes in its accelerations is described by the Langevin equation, given by

$$m\frac{du}{dt} = -fu + F(t), \tag{4.1}$$

where u is the velocity of the particle and m is the mass. On the right-hand side is written the influence of the surrounding medium, composed of two parts: a fluctuating force $F(t)$, and a deterministic part $-fu$ representing dynamical friction. Or, equivalently, we have

$$\frac{du}{dt} = -\beta u + A(t), \tag{4.2}$$

where $\beta = f/m$ and $A(t) = F(t)/m$. The analogous electrical problem is an L–R circuit described by the equation

$$L\frac{di}{dt} + Ri = E(t), \tag{4.3}$$

where $L(t)$ is a purely fluctuating electromotive force (a thermal noise source). Certain crucial assumptions are involved in the solution of this problem:

(i) The mean of the fluctuating force $A(t)$ over the ensemble of particles starting with the same initial velocity u_0 at $t = 0$ is zero:

$$\mathrm{E}\{A(t)\} = 0. \tag{4.4}$$

(ii) The values of $A(t)$ at two different times t_1 and t_2 are not correlated at all except for small intervals $(t_2 - t_1)$. More precisely,

$$\mathrm{E}\{A(t_1)A(t_2)\} = \phi(|t_1 - t_2|) \tag{4.5}$$

and $\phi(x)$ is peaked around the zero value of its argument.

(iii) The correlations of $A(t)$ obey the conditions

$$\mathrm{E}\{A(t_1)A(t_2)\ldots A(t_{2n+1})\} = 0,$$
$$\mathrm{E}\{A(t_1)A(t_2)\ldots A(t_{2n})\} = \sum_{\text{all pairs}} \mathrm{E}\{A(t_i)A(t_j)\}\mathrm{E}\{A(t_k)A(t_l)\}\ldots, \tag{4.6}$$

so that $A(t)$ is a purely Gaussian white noise.

Since we are interested in writing an equation for the probability $\pi(u; t)$ that the velocity of the particle lies between u and $u + du$ at time t, we tacitly assume that u is a Markov process. This may be warranted by the physical features of the theory.

In order to arrive at a solution for the Fokker–Planck equation governing $\pi(u; t)$ it is necessary to assume that $A(t)$ is independent of u, and that $A(t)$ varies very rapidly compared to u, implying that $A(t)$ may undergo very many fluctuations in a short interval Δ during which $u(t)$ varies only slightly. This is a consequence of the fact that m, the mass of the particle, is much larger than that of the molecules that surround it. The drastic nature of these assumptions is discussed a little later.

Integrating (4.2), we obtain

$$u = u_0 e^{-\beta t} + e^{-\beta t}\int_0^t e^{\beta \xi} A(\xi)\, d\xi. \tag{4.7}$$

The mean and mean square values of $u(t)$ are given by

$$\mathbf{E}\{u(t)\} = u_0 e^{-\beta t},$$

$$\mathbf{E}\{u^2(t)\} = u_0{}^2 e^{-2\beta t} + e^{-2\beta t} \int_0^t \int_0^t e^{\beta(x+y)} \mathbf{E}\{A(x)A(y)\}\, dx\, dy$$

$$= u_0{}^2 e^{-2\beta t} + \frac{l}{2\beta}(1 - e^{-2\beta t}), \tag{4.8}$$

where $l = \int_{-\infty}^{+\infty} \phi(w)\, dw$. By the theorem of equipartition of energy which is to prevail in the medium in equilibrium, we have

$$\lim_{t \to \infty} \mathbf{E}\{u^2(t)\} = \frac{l}{2\beta} = \frac{kT}{m}, \tag{4.9}$$

where k is the Boltzmann constant and T the temperature at equilibrium. Hence we have the mean square velocity at time t, given by

$$\mathbf{E}\{u^2(t)\} = \frac{kT}{m} + \left(u_0{}^2 - \frac{kT}{m}\right) e^{-2\beta t}, \tag{4.10}$$

an equation expressing the approach to equilibrium.

We can consider the problem of evaluating the mean and mean square displacement of the free particle in Brownian motion which, at time $t = 0$, starts from $x = x_0$ with velocity u_0. Integrating (4.7) over time, we obtain

$$x = x_0 + \frac{u_0}{\beta}(1 - e^{-\beta t}) + \int_0^t e^{-\beta \eta}\, d\eta \int_0^{\eta} e^{\beta \xi} A(\xi)\, d\xi, \tag{4.11}$$

from which we obtain, for the average value of the displacement at time t,

$$\mathbf{E}\{x(t) - x_0\} = \frac{u_0}{\beta}(1 - e^{-\beta t}), \tag{4.12}$$

which corresponds to the distance traveled in time t with a velocity $u_0 e^{-\beta \tau}$. After some calculation, we arrive at the mean square average of distance traveled as

$$\mathbf{E}\{[x(t) - x_0]^2\} = \frac{l}{\beta^2} t + \frac{u_0{}^2}{\beta}(1 - e^{-\beta t})^2 + \frac{l}{2\beta^2}(-3 + 4e^{-\beta t} - e^{-2\beta t}), \tag{4.13}$$

so that for long intervals of time we have

$$\mathbf{E}\{[x(t) - x_0]^2\} = \frac{l}{\beta^2} t = \frac{2kT}{m\beta} t. \tag{4.14}$$

This result is the same as that obtained by Einstein which led to the determination of Avogadro's number. It can be shown that $u - u_0 e^{-\beta t}$ obtained from (4.11), as well as $x(t) - \mathbf{E}\{x(t)\}$ evaluated from (4.11), has odd moments zero and even moments related to the second moment exactly as the even moments of a Gaussian distribution. This implies the probability frequency functions $\pi(u|u_0; t)$ and $F(x|x_0; t)$ corresponding to the velocity u, and the displacements x of the Brownian particle are given by

$$\pi(u|u_0; t) = \left(\frac{m}{2kT(1 - e^{-\beta t})}\right)^{\frac{1}{2}} \exp\left\{\frac{m}{2kT} \frac{(u - u_0 e^{-\beta t})^2}{(1 - e^{-\beta t})}\right\}, \tag{4.15}$$

$$F(x|x_0; t) = \left[\frac{m}{2\pi kT(2\beta t - 3 + 4e^{-\beta t} - e^{-2\beta t})}\right]^{\frac{1}{2}} \cdot$$

$$\exp\left\{\frac{m\beta^2}{2kT} \frac{[x - x_0 - u_0(1 - e^{-\beta t})]^2}{2\beta t - 3 + 4e^{-\beta t} - e^{-2\beta t}}\right\}. \tag{4.16}$$

4.2 Fokker–Planck Description

We now obtain the Fokker–Planck equation for the probability frequency function $\pi(u; t)$, starting from the Langevin equation. From the Langevin equation we can calculate the moments of the increments of velocity which are necessary to arrive at the Fokker–Planck equation of the velocity distribution. If we denote by Δu the change in u for a small interval of time Δt, we have

$$\Delta u = - \beta u \Delta t + \int_t^{t + \Delta t} A(\xi) \, d\xi. \tag{4.17}$$

We assume that the relaxation time Δt for the Brownian particle is very large compared with the time in which fluctuations occur in the random force. For the solution of the problem this assumption is crucial, and it decouples statistically the fluctuating accelerations imparted to the particle by the random hittings from the velocity of the particle at any time. The first two moments of Δu are given by

$$E\{\Delta u\} = - \beta u \Delta t, \tag{4.18}$$

$$E\{(\Delta u)^2\} = \int_t^{t + \Delta t} \int_t^{t + \Delta t} E\{A(\xi_1) A(\xi_2)\} \, d\xi_1 \, d\xi_2. \tag{4.19}$$

It is easy to see that higher moments of $(\Delta u)^n$ $(n > 2)$ are of higher order in Δt. Using the technique outlined in Section 2.4 of Chapter 3, we obtain

$$\frac{\partial \pi(u; t)}{\partial t} = \beta \frac{\partial(u\pi(u; t))}{\partial u} + D \frac{\partial^2 \pi(u; t)}{\partial u^2}. \tag{4.20}$$

To solve this equation we introduce the variables $\tau = t\beta$ and $y = u(\beta/D)^{\frac{1}{2}}$ to obtain

$$\frac{\partial \pi(y; \tau)}{\partial \tau} = \frac{\partial^2 \pi(y; \tau)}{\partial y^2} + y \frac{\partial \pi(y; \tau)}{\partial y} + \pi(y; \tau) \tag{4.21}$$

with the initial condition $\pi(y; 0) = s(y)$ and boundary condition $\pi(t; y) = 0$ when $y = \pm \infty$. Equation (4.21) can be solved by separation of variables:

$$\pi(y; \tau) = \sum A_n e^{-n\tau} H_n(y) e^{-y^2/4} \tag{4.22}$$

with

$$H_n(y) = (-1)^n e^{-y^2/4} \frac{d^n}{dy^n} e^{y^2/2}.$$

The constants A_n's are determined by the use of the initial condition

$$s(y) = \sum_{n=0}^{\infty} A_n H_n(y) e^{-y^2/4}. \tag{4.23}$$

Thus we have

$$A_n = \frac{1}{(2\pi)^{\frac{1}{2}} n!} \int_{-\infty}^{+\infty} H_n(\lambda) s(\lambda) e^{-\lambda^2/4} \, d\lambda. \tag{4.24}$$

The solution can be written in the form

$$\pi(u; t) = \frac{1}{(2\pi)^{\frac{1}{2}}} \int_{-\infty}^{+\infty} \left[\sum_{n=0}^{\infty} \frac{H_n(y) H_n(\lambda)}{n!} \cdot e^{-n\tau} e^{-(\lambda^2+y^2)/4} s(\lambda) \right] d\lambda. \tag{4.25}$$

If we use the identity

$$\sum_{n=0}^{\infty} \frac{H_n(y) H_n(\lambda)}{n!} e^{-n\tau} =$$

$$\int \frac{e^{(y^2+\lambda^2)/4}}{(1-e^{-2\tau})^{\frac{1}{2}}} \exp\left\{ -\frac{(y^2 + \lambda^2 e^{-2\tau} - 2y\lambda e^{-\tau})}{2(1-e^{-2\tau})} \right\} s(\lambda) \, d\lambda, \tag{4.26}$$

we obtain for the solution of (4.21), with the initial condition $s(y) = \delta(y - y_0)$,

$$\pi(y; t) = \frac{1}{2\pi(1-e^{-2\tau})^{\frac{1}{2}}} \exp\left\{ -\frac{(y - y_0 e^{-\tau})^2}{2(1-e^{-2\tau})} \right\}, \tag{4.27}$$

which in terms of u, u_0, and t transforms to the well-known Uhlenbeck–Ornstein solution given by

$$\left[\frac{m}{2kT(1-e^{-2\beta t})} \right]^{\frac{1}{2}} \exp\left\{ -\frac{m(x - x_0 e^{-\beta t})^2}{2kT(1-e^{-2\beta t})} \right\}. \tag{4.28}$$

4.3. Invariant Imbedding Equation

Starting from the Langevin equation, we can apply the invariant imbedding method of Bellman and Harris [12] (see also [13]) to obtain a differential equation for $\pi(u|u_0, t)$ by considering what happens to u in the first interval Δt of the time axis. The loss of velocity due to dynamical function during this interval, in view of the Langevin equation, is $-\beta u_0 \Delta t$. The gain in velocity due to random accelerations resulting from the external forces under the assumptions mentioned and explained earlier is $M(\Delta t) = \int_0^{\Delta t} A(\xi)\, d\xi$. This is a fluctuating quantity and is expected to occur with probability $\phi(M_0)$. Hence we set

$$\pi(u_1|u_0, t, 0) = \int \pi(u|u_0 - \beta u_0 \Delta t + M(\Delta t); (t - \Delta t)\phi(M(\Delta t))\, dM \quad (4.29)$$

with

$$\int \phi(M)\, dM = 1, \qquad \int M\phi[M(\Delta t)]\, dM = 0,$$

$$\int M^2 \phi[M(\Delta t)]\, dM = 2D\Delta t.$$

As Δt tends to zero, we obtain

$$\frac{\partial \pi(u|u_0; t)}{\partial t} = \beta u_0 \frac{\partial \pi(u|u_0; t)}{\partial u_0} - D\frac{\partial^2}{\partial u_0^2}\pi(u|u_0; t). \quad (4.30)$$

5. DISCUSSION OF BASIC ASSUMPTIONS

In this section we examine critically the assumptions leading to the setting up of the Fokker–Planck equation and consider other possible types of fluctuating forces.

5.1. Markov Property

In all the investigations relating to the Langevin equation it is generally assumed that the random process represented by the velocity distribution of the Brownian particle is a Markov process and that the fluctuating force and the velocity are independent owing to the relatively large relaxation time of the Brownian particle compared to the fluctuation time of the random forces. In addition, the first two moments of the increments in velocity are taken to be proportional to the interval Δt, the higher moments being of higher order in Δt. Though these assumptions are physically valid, we can examine the different contexts in which they can be violated.

It is interesting to point out that the assumption about the moments need not hold if the transition probability (see Lax [14]; see also Srinivasan and Vasudevan [15]) is of the form

$$\pi(u'|u; \Delta t) = R(u'|u)\, \Delta t + \delta(u - u')\{1 - \Delta t \int R(u'|u)\, du'\}. \quad (5.1)$$

This is also the case in the model of the fluctuating density fields investigated by Ramakrishnan [16]. In this case all the moments are proportional to Δt, and hence the Fokker–Planck equation will be the generalized one (see [15]):

$$\frac{\partial P(u|u_0; t)}{\partial t} = \sum_{m=1}^{\infty} \frac{1}{m!}\left(\frac{\partial}{\partial u}\right)^m [P(u|u_0; t)a_m(u)] \tag{5.2}$$

with

$$a_m = \int (u' - u)P(u'|u; \Delta t)\, du'.$$

In the light of the viewpoint expressed in Section 2.4 of Chapter 3, it is in order to investigate whether the Langevin equation with the assumptions of the fluctuating force F does imply the Markov property of u. To this end we consider the vector process $\pi(u, F|u_0, F_0; t)$ which is certainly Markovian, and we can write the Chapman–Kolomogorov equation

$$\pi(u, F|u_0, F_0; t + \Delta) = \iint \pi(u', F'|u_0, F_0; t)\pi(u, F|u', F'; \Delta t)\, du'\, dF'. \tag{5.3}$$

By the usual procedure we can obtain the generalized Fokker–Planck equation,

$$\frac{\partial \pi}{\partial t} = \sum_{n=1}^{\infty} \frac{1}{n!} \sum_{r=0}^{n} \binom{n}{r}\frac{\partial^n}{\partial u^{n-r}\, \partial F^r}(\alpha'_{n-r,r}\, \pi) \tag{5.4}$$

with

$$\alpha'_{m,n} = \iint (u' - u)^m (F' - F)^n \pi(u', F'|u_0, F_0; \Delta t)\, du'\, dF'.$$

If we note that

$$\int \pi(u', F'|u, F; \Delta t)\, du' = p(F'|F, \Delta t) \tag{5.5}$$

and assume that

$$P(F'|F, \Delta t) = \frac{1}{(4\pi D\Delta t)^{\frac{1}{2}}} \exp\left\{ -\frac{(F - F')^2}{2D\Delta t} \right\} \tag{5.6}$$

with other necessary assumptions on F, we can still recover the usual Fokker–Planck equation as explained in [15] provided $p(F|F', \Delta)$ is given by (5.6).

In the Langevin equations we can replace the random force by a wildly fluctuating density field formulated by Ramakrishnan [16]. This is governed by a transition probability similar to (5.1), and the integral of this field over time can possess moments similar to that of a Gaussian process. This can be a good description of the random force on the right-hand side of the Langevin equation (4.1) and can lead to the usual Fokker–Planck equation with suitable assumptions.

5.2. Fluctuation and Dissipation

In the Langevin equation, $F(t)$ was assumed to be a Gaussian process with a delta correlation. Then, if we solve the Fokker–Planck equation and impose the condition that the Brownian particle when in thermal equilibrium in the medium must attain Maxwell distribution, we have

$$\pi(u, t|u_0, t_0) = c \exp\left(-\tfrac{1}{2} \frac{mu^2}{kT} \right). \tag{5.7}$$

This requires a definite relation between the diffusion coefficient and the functional constant. Thus the correlations in the fluctuating force and the frictional force, which is deterministic, in nature are connected with each other intrinsically by

$$m\gamma = \frac{m^2 D}{kT} = \frac{1}{kT} \int_0^\infty \mathrm{E}\{F(t_0)F(t_0 + t)\}\, dt. \tag{5.8}$$

Putting this in different language, the power spectrum of the random force which is white in the present case, is given by

$$G_F = \frac{mkT}{\pi} \gamma. \tag{5.9}$$

Thus we can say that the power spectrum of the random force is determined by the frictional force. In general, the resistance in a given system represents the manner by which external work is dissipated into microscopic thermal energy. The reverse process is the generation of the random force as a result of thermal fluctuations. This is the essence of the Nyquist relations [17] obtained by thermodynamic reasoning implying that the fluctuating voltage across a resistor is determined by its impedance.

A generalization of the Brownian motion to the case when the particle is not necessarily heavier than the surrounding medium particles which are interacting with it is a very interesting piece of work due to Mori [18] and Kubo [19]. Such a generalization implies that the assumption that the relaxation time of the random force is very small compared to the relaxation time of the particle has to be abandoned. This assumption yields a correlation function in equilibrium

$$\frac{d}{dt}\mathrm{E}\{u(t_0)u(t_0 + t)\} = -\gamma\mathrm{E}\{u(t_0)u(t_0 + t)\}. \tag{5.10}$$

If, however, we want the correlation in a stationary process to satisfy the relation

$$\frac{d}{dt_0}\mathrm{E}\{u(t_0)u(t_0 + t)\} = 0, \tag{5.11}$$

Mori [18] and Kubo [19] have shown that the assumption of constancy of γ, has to be changed. Equation (5.10) is no longer valid for small time intervals t of the Brownian particle. To resolve this difficulty, they extended the Langevin equation by including a retarded effect in the friction term. The generalized Langevin equation is of the form

$$\frac{du(t)}{dt} = - \int \gamma(t - t')u(t')\, dt' + \frac{1}{m}F(t) \tag{5.12}$$

with the assumptions

$$\begin{aligned} E\{F(t)\} &= 0; \\ E\{u(t_0)F(t)\} &= 0, \qquad t > t_0. \end{aligned} \tag{5.13}$$

We have in the equilibrium state

$$m\gamma(\omega) = \frac{1}{RT}\int E\{F(t_0)F(t_0 + t)\}e^{-i\omega t}\, dt, \tag{5.14}$$

where $\gamma(\omega)$ is the Fourier transform of the friction $\gamma(t)$. This is a relation for the noise in a resistor with a frequency-dependent impedance. It can be a generalized version of the Nyquist theorem. It has been shown by Mori [18] that the equations of motion can generally be transformed into the form of the generalized Langevin equation in a formal manner. The generalized Nyquist relation and its relevance in statistical mechanics of irreversible systems have been dealt with by Mori [18] in great detail.

The importance of these extensions is felt in problems like the motion of a very heavy isotropic impurity in an infinite harmonic lattice with nearest neighbor interactions. In 1960, Rubin [20] studied this problem in detail and found that for a one-dimensional lattice the process corresponds to the Brownian motion. Later, Mazur and Braun [21] and Ford, Kac, and Mazur [22] were able to derive a generalized Langevin type of equation for the impurity particle without the restriction of the nearest neighbor interactions. When the frequency spectrum of the lattice satisfies some weak conditions, it goes into the ordinary Langevin equation for a free Brownian particle with Gaussian random force of a purely random nature in the limit of infinite impurity mass. For finite mass, the theory of Mori [18] is applicable; the reader is referred to the paper by Nakazawa [23] for further details.

6. SCHRODINGER EQUATION

There have been several attempts to link the Schrodinger equation with the equation of a Brownian particle subjected to random forces, and vice versa. These efforts spring from the idea of giving an alternative interpretation to quantum theory by assuming that the Schrodinger field arises as an average over random fluctuations at a subquantum level and is related to the ideas of

Louis de Broglie [24], Madeling [25], and Bohm [26, 27]. We describe below some of the recent efforts of Auerback, Braun, and Garcia Collin [28, 29] to show that the motion of a Brownian particle under Smoluchowsky approximation can be described by a Schrodinger-like equation defining a complex amplitude whose norm is the same as the probability density.

Let $x(t)$ be the stochastic process and $\rho(x, t)$ the probability density at (x, t). The conservation of density yields (for one-dimensional processes)

$$\frac{\partial \rho}{\partial t} + \operatorname{div}(v\rho) = 0, \tag{6.1}$$

where v is the macroscopic flow velocity given in general by

$$v = a + \frac{1}{\rho}\frac{\partial}{\partial x}(v\rho), \tag{6.2}$$

where $a = K/\beta$, K being the external force per unit mass and β the frictional coefficient. The second term in (6.2) refers to diffusion. Because ρ is a positive single-valued function, we set

$$\rho = e^{2R} \tag{6.3}$$

and obtain

$$\frac{\partial R}{\partial t} = -\tfrac{1}{2}\frac{\partial v}{\partial x} - v \operatorname{grad} R. \tag{6.4}$$

If we introduce the function ψ such that $\psi\psi^* = \rho$, obeying the situation described by (6.1), and assume that $\psi = e^{R+iS}$, it can be shown (for S real) that (6.1) can be cast in the form

$$i\frac{\partial \psi}{\partial t} = -\tfrac{1}{2}\alpha\nabla^2\psi + V\psi, \tag{6.5}$$

where $v = \alpha \operatorname{grad} S$, corresponding to an irrotational flow in configurational space. But the potential in the Schrodinger-like equation in (6.5) is given by

$$V = -\frac{\partial S}{\partial t} + \tfrac{1}{2}\alpha[\nabla^2 R + (\operatorname{grad} R)^2 - (\operatorname{grad} S)^2]. \tag{6.6}$$

The physical meaning of this potential is investigated in detail in [27] and [28]. It is interesting to point out that, if we put the external force equal to zero after some laborious calculations, we can show that V is proportional to v^2. This indicates that, even in the absence of the external force or frictional force, there is an interaction proportional to v^2. In this case the potential may be interpreted as a simple Rayleigh dissipation function. A more interesting program is to start from the Schrodinger equation,

$$i\hbar\frac{\partial \psi}{\partial t} = -\frac{\hbar^2}{2m}\frac{\partial^2 \psi}{\partial x^2} + V\psi, \tag{6.7}$$

and to transform it into a Smoluchowsky equation for the probability density ρ which is the modulus square of the amplitude ψ. If we write $\psi = e^{R+iS}$, we can obtain the systems of equations

$$\frac{\partial R}{\partial t} = -\tfrac{1}{2}\alpha\frac{\partial^2 S}{\partial x^2} - \alpha\frac{\partial R}{\partial x}\frac{\partial S}{\partial x}, \qquad (6.8)$$

$$-\frac{\partial S}{\partial t} = -\tfrac{1}{2}\alpha\frac{\partial^2 R}{\partial x^2} - \tfrac{1}{2}\alpha\left[\left(\frac{\partial R}{\partial x}\right)^2 - \left(\frac{\partial S}{\partial x}\right)^2\right] + \frac{V}{h}, \qquad (6.9)$$

where $\alpha = \hbar/m$. From (6.8), we have the continuity equation,

$$\frac{\partial \rho}{\partial t} + \frac{\partial}{\partial x}\left[\alpha\rho\frac{\partial S}{\partial x}\right] = 0 \qquad (6.10)$$

with

$$\rho = e^{2R} = \psi\psi^*.$$

The other equation for S can be written

$$\frac{\partial S}{\partial t} + \left(\frac{\partial}{\partial x}\frac{\hbar s}{2m}\right)^2 + V - \frac{\hbar^2}{2m}\left[\frac{\partial^2 R}{\partial x^2} + \left(\frac{\partial R}{\partial x}\right)^2\right] = 0, \qquad (6.11)$$

which is called by Bohm [27] the Hamilton–Jacobi equation for the action $\hbar S$. In addition to the classical potential V, we have an extra term called the Bohm potential,

$$V_B = -\frac{\hbar^2}{2m}\left[\frac{\partial^2 R}{\partial x^2} + \left(\frac{\partial R}{\partial x}\right)^2\right]. \qquad (6.12)$$

To arrive at the equation of Brownian motion, we introduce

$$Q = R + S \qquad (6.13)$$

and hence rewrite (6.8) as

$$\frac{\partial R}{\partial t} = -\tfrac{1}{2}\alpha\left[\frac{\partial^2 Q}{\partial x^2} - \frac{\partial^2 R}{\partial x^2}\right] - \alpha\frac{\partial R}{\partial x}\left(\frac{\partial Q}{\partial x} - \frac{\partial R}{\partial x}\right), \qquad (6.14)$$

which can be easily transformed into

$$\frac{\partial \rho}{\partial t} + \frac{\partial}{\partial x}\left[\left(\alpha\frac{\partial Q}{\partial x}\right)\rho - \tfrac{1}{2}\alpha\frac{\partial \rho}{\partial x}\right] = 0. \qquad (6.15)$$

This is the well-known Smoluchowsky equation for Brownian motion, acted on by an external force K per unit mass given by

$$K = \alpha\beta\frac{\partial Q}{\partial x} \qquad (6.16)$$

with the usual diffusion coefficient given by

$$D = \frac{\alpha}{2} = \frac{\hbar}{2m}. \qquad (6.17)$$

Following this interpretation, we shall say that β is a parameter which measures the interaction of the particle with the vacuum. The fluctuation in the stochastic field acting on the particle is proportional to $\hbar$. However, it is important to note that, though the Schrodinger equation is valid for all times, the Fokker–Planck equation (6.15) is valid only for time intervals $\beta \Delta t \gg 1$. The description of the motion of a classical Brownian particle is valid for time intervals greater than the relaxation time, that is, for $\Delta t \gg \beta^{-1}$. Owing to stochastic nature of the problem, there is a mean square deviation of position and velocity coordinates. A calculation of these deviations from the solution of Smoluchowsky equation leads to uncertainty relations. Proceeding further, we can also establish the formal analogy between quantum mechanical equations and those of hydrodynamics. It is also shown that, in the case of the harmonic oscillator, the external force term contains a part proportional to elastic forces and another part relating to the interaction of the system with the vacuum which directly leads to quantization of the system. Thus we come to the conclusion that, even with all the restrictions implied in a stochastic interpretation of quantum mechanical equations of motion, we can obtain some intuitive understanding of the underlying processes, and pushing forward this program is not without sufficient promise.

Before concluding this section, we want to point out that Brownian particle trajectories play an important role in the formulation of Wiener integrals introduced in the 'twenties and worked out by a number of mathematicians in the 'fifties. Later Feynman initiated the path integration formalism in quantum mechanics which is similar to Wiener integrals in classical physics. Details of these developments are dealt with in Chapter 8.

REFERENCES

1. R. Brown, *Philos. Mag.*, 2nd Ser., **4**(1828), 161.
2. A. Einstein, *Ann. der Physik*, **17**(1905), 549; **19**(1906), 371.
3. N. Wiener, Generalised Harmonic Analysis, *Acta Math.*, **55**(1930), 117.
4. S. Chandrasekhar, *Rev. Mod. Phys.*, **15**(1943), 1.
5. Alladi Ramakrishnan, in *Handbuch der Physik*, Vol. 3, Springer-Verlag, Berlin, 1959.
6. G. E. Uhlenbeck and L. S. Ornstein, *Phys. Rev.*, **36**(1930), 823.
7. M. C. Wang and G. E. Uhlenbeck, *Rev. Mod. Phys.*, **17**(1945), 327.
8. R. L. Stratonovich, *Topics in the Theory of Random Noise*, Vol. I, R. A. Silverman, Gordon and Breach, New York, 1963.
9. W. Feller, *Ann. Math.*, **55**(1952), 468.
10. E. Hille and R. S. Phillips, *Functional Analysis and Semi-groups*, Amer. Math. Soc., Providence, R.I., 1957.

11. K. Yosida, *Proc. Int. Cong. Math.*, *Amsterdam*, (1957), 405.
12. R. E. Bellman and T. E. Harris, *Proc. Natl Acad. Sci. U.S.*, **34**(1948), 601.
13. R. E. Bellman, R. Kalaba, and G. M. Wing, *J. Math. Phys.*, **1**(1960), 280.
14. M. Lax, *Rev. Mod. Phys.*, **32**(1960), 25.
15. S. K. Srinivasan and R. Vasudevan, *Ann. Inst. Henri Poincaré*, Sect. A, **7**(1967), 303.
16. Alladi Ramakrishnan, *Astrophys. J.*, **119**(1954), 443.
17. H. Nyquist, *Phys. Rev.*, **32**(1928), 110.
18. A. Mori, *J. Phys. Soc. Japan*, **33**(1965), 423.
19. R. Kubo, *J. Phys. Soc. Japan*, **12**(1957), 520.
20. R. J. Rubin, *Proc. Int. Symp. Transport Processes in Statist. Mechanics*, I. Prigogine, ed., Interscience, New York, 1958.
21. P. Mazur and E. Braun, *Physica*, **30**(1964), 1973.
22. G. W. Ford, M. Kac, and P. Mazur, *J. Math. Phys.*, **6**(1965), 504.
23. H. Nakazawa, *Progr. Theoret. Phys.*, *Suppl.*, **36**(1966), 172.
24. Louis de Broglie, *Une Tentative d'Interprétation Causale et non Linéaire de la Méchanique Ondulatoire*, Gauthiers-Villars, Paris, 1956, translated by A. J. Knodell and Jack C. Miller, Elsevier, Amsterdam, 1960.
25. F. Madeling, *Z. Physik*, **40**(1926), 332.
26. D. Bohm, *Causality and Chance in Modern Physics*, Van Nostrand, Princeton, N.J. 1957.
27. D. Bohm, *Phys. Rev.*, **85**(1952), 166.
28. L. dela Pena Auerbach, E. Braun, and L. S. Garcia Collin, *J. Math. Phys.*, **9**(1968), 668.
29. L. dela Pena Auerbach and L. S. Garcia Collin, *J. Math. Phys.*, **9**(1968), 916.

Chapter 5

RESPONSE PHENOMENA

1. INTRODUCTION

In the previous chapter we discussed in detail the differential equation governing the Brownian motion. The physical quantities of interest in the case of Brownian motion are the velocity and displacement suffered by the particles. If we neglect the nonlinear effects, the displacement x (or the velocity) can be expressed as the weighted integral over time of the random force $F(t)$. In the light of the discussion in Section 5 of Chapter 4, the random force $F(t)$, which is usually Gaussian in nature, can be related to an appropriately weighted Poisson process (more generally, a point process). Such a viewpoint enables us to consider the displacement as the cumulative response to the point process in question so that the solution of a class of linear stochastic differential equations can be discussed from the standpoint of response phenomena. The objective of this chapter is to study some of the physical processes which fall in this category.

In Section 2 we discuss the phenomenon of shot noise, taking into account the space charge effects. Next we deal with the Barkhausen noise in the light of certain non-Markovian models. Section 4 is devoted to the statistical properties of fluctuating light beams, a topic expected to be of considerable importance especially in view of the current research in laser beams.

2. SHOT NOISE

The phenomenon of shot noise is historically one of the earliest examples of the integral of a random function. The theory of shot effect (see Chapter 2, Section 2) originally anticipated by Campbell [1] and subsequently developed by Rice [2] gives an adequate description of the noise current when the system has reached equilibrium. The generalization to the nonequilibrium situation was done by Moyal [3], who discussed the voltage fluctuations arising from shot noise. In all these treatments of shot noise, it has been recognized that the shot effect consists of a series of pulses of currents distributed in a Poisson manner (1 electron = 1 pulse) and that we can observe only the response of the circuit to the current pulses and not the pulses themselves. If $\phi(t)$ (as a function of time) is the response to a single pulse at time t after the occurrence of the pulse (at $t = 0$), $\psi(t)$, the cumulative response due to the series of pulses that have occurred up to t, is given by

$$\psi(t) = \sum_i \phi(t - t_i)H(t - t_i), \qquad (2.1)$$

where $H(x)$ is the Heaviside unit function. If we consider $V(t)$ the voltage in the anode circuit with inductance L, resistance R, and capacitance C, due to the arrival of an electron at $t = 0$, $V(t)$ is given by (see, for example, Rowland [4])

$$V(t) = \tfrac{1}{2}\left(-\frac{\varepsilon}{C} \right)\left[\left(1 + \frac{R}{2i\omega L} \right) \exp -\left(\frac{R}{L} - i\omega \right)t + \right.$$

$$\left. \left(1 - \frac{R}{2i\omega L} \right) \exp -\left(\frac{R}{L} + i\omega \right)t \right], \qquad t > 0$$

$$= 0, \qquad t < 0, \qquad (2.2)$$

where ω is given by

$$\omega = \left(\frac{1}{CL} - \frac{R^2}{4L^2} \right)^{\tfrac{1}{2}}. \qquad (2.3)$$

If $W(v)$ is the Fourier transform of $V(t)$, the Fourier transform of $\phi(t)$ is $W(v)G(v)$, where $G(v)$ is the amplification factor that is set for the frequency $v/2\pi$.

Thus a study of $\psi(t)$, or rather the probability distribution of $\psi(t)$, leads us toward a good understanding of the phenomenon of shot noise. In fact, the Campbell's theorem introduced in Section 2 of Chapter 2 and its generalizations (see Rice [2]) explain the properties of the moments of $r(t)$. Such a mode of description is only an approximation and, in fact, Hull and Williams [5] found that the measured shot voltage fell below even 40 percent of the theoretical value. Of course, this was attributed mainly to the fluctuations in voltage, due to the space charge effect, which alter the number density of arrival of electrons at the anode. Meanwhile, Johnson [6], who made some measurements of shot voltage in space-charge-limited tubes, pointed out that the expected value of shot voltage could be calculated as for temperature-limited currents by assuming an internal resistance of the valve. Using this suggestion as well as the experimental findings of Moullin [7], Rowland [8] formulated the problem in a precise form. He assumed that electrons might, with specific probabilities, have lives of any given length on the anode system during which time each of them adds $(- \varepsilon/C)\exp [- (t - t_i)/RC]$ to the anode potential of the valve whose anode-to-earth capacity is C and feed resistance is R. In addition, there is an effective internal resistance ρ of the valve, determined by the assumption that a variation of the anode potential causes a variation $1/\varepsilon\rho$ times as great, in the probable density of arrival of electrons. Thus the arrival of each electron at time t_i decreases the probable density of further arrivals by an amount $(1/C\rho)\exp[-(t-t_i)/CR]$. If N_0 is the density of arrivals at time $t = 0$, then $N(t)$, the arrival density at any time $t > 0$, is given by

$$N(t) = N_0 - \sum_i \alpha(t - t_i)N_0 - \sum \alpha(t - t_1) > 0 \qquad (2.4)$$

$$= 0 \qquad \text{otherwise,}$$

where

$$\alpha(t - t_i) = \left(\frac{1}{C\rho}\right)\exp\left[-\frac{(t - t_i)}{CR}\right]. \qquad (2.5)$$

With the modification of the density of arrivals as given by (2.4), it is clear that the probability frequency function governing the number of arrivals in the interval $(0, t)$ is no longer Poissonian in nature. On the other hand, the times of arrivals constitute a non-Markovian point process on the time axis. It is eminently reasonable to regard the point process as a generalized inhomogeneous Poisson process. Of course, we could see from (2.4) that the Poisson nature is completely destroyed, the parameter $\lambda(t)$ being a stochastic variable* with λ_0 as its initial value. $\lambda(t)$ satisfies the differential equation

$$\frac{d\lambda}{dt} = a(\lambda_0 - \lambda) - b\frac{dn(t)}{dt}, \qquad a = \frac{1}{CR}, \qquad b = \frac{1}{C\rho}, \qquad (2.6)$$

where $n(t)$ represents the number of arrivals in the interval $(0, t)$. The cumulative response $\psi(t)$ can be expressed as the stochastic integral

$$\psi(t) = \int_0^t dn(\tau)\phi(t - \tau), \qquad (2.7)$$

where $\phi(t)$ is the response due to any individual electron arriving at $t = 0$. A complete study of the stochastic differential equation (2.6) will enable us to obtain the statistical properties of $\psi(t)$.

In view of the non-Markovian structure of the point process implied by (2.6), it is very difficult to obtain the p.f.f. of $\psi(t)$. An analysis of the point process leading to the correlations of $\psi(t)$ of the first few orders has been given by Srinivasan [11]. In the next few subsections, we present an account of the process and obtain an explicit expression for the power spectrum of the cumulative response.

* See Srinivasan [9]. Bartlett [10] has also considered such processes in connection with spectral analysis of point processes. He calls such processes doubly stochastic.

2.1. Probability Frequency Function of $\lambda(t)$

We shall assume that the process is switched on at $t = 0$, when the probability of the occurrence of an event in the infinitesimal $(0, \Delta)$ is λ_0 (λ_0 being a constant). Thus the first event happens between t_1 and $t_1 + dt_1$ with probability $\exp[- \lambda_0 t_1]\lambda_0 \, dt_1$, while the probability of occurrence of the second event between t_2 and $t_2 + dt_2$ is given by

$$P_2(t_1, t_2) \, dt_2 = \exp\left\{ - \int_{t_1}^{t_2} [\lambda_0 - b \exp - a(t' - t_1)] \, dt' \right\} \cdot$$

$$\{\lambda_0 - b \exp - a(t_2 - t_1)\} \, dt_2. \qquad (2.8)$$

In analogy with the inhomogeneous Poisson process, let us denote the parameter characterizing the process by $\lambda(t)$. The parameter $\lambda(t)$ is no longer a deterministic function of t but depends on the various random values of t at which the events have occurred. A typical realized values of $\lambda(t)$ corresponding to the events that have occurred at $t_1, t_2, \ldots, t_n$ is given by

$$\lambda^R(t) = \lambda_0 - b \sum_{i=1}^{n} \exp - a(t - t_i). \qquad (2.9)$$

The probability measure corresponding to the preceding realized value can be calculated using (2.8).

Let $\pi(\lambda, t)$ be the probability frequency function of $\lambda(t)$ so that $\pi(\lambda, t) \, d\lambda$ denotes the probability that $\lambda(t)$ has a value between λ and $\lambda + d\lambda$ at t. Let us increase t by Δ. If an event does not occur between t and $t + \Delta$, $\lambda(t)$ increases deterministically during the interval $(t, t + \Delta)$, as is evident from (2.6), the rate of increase being given by

$$\frac{d\lambda}{dt} = a(\lambda_0 - \lambda). \qquad (2.10)$$

If, on the other hand, an event occurs between t and $t + \Delta$, λ suddenly diminishes by b. Using these results, we obtain

$$\pi(\lambda, t + \Delta) \, d\lambda = (1 - \lambda\Delta)\pi(\lambda - a\overline{\lambda_0 - \lambda}\Delta, t)d[\lambda - (\lambda_0 - \lambda)a\Delta] +$$
$$\pi(\lambda + b, t)(\lambda + b)\Delta \, d\lambda + o(\Delta). \qquad (2.11)$$

Proceeding to the limit as Δ tends to zero, we obtain

$$\frac{\partial \pi(\lambda, t)}{\partial t} = (a - \lambda)\pi(\lambda, t) - a(\lambda_0 - \lambda)\frac{\partial \pi(\lambda, t)}{\partial \lambda} + (\lambda + b)\pi(\lambda + b, t). \qquad (2.12)$$

This equation is true for $\lambda > 0$. When $\lambda < 0$, $\pi(\lambda, t)$ satisfies the equation

$$\frac{\partial \pi(\lambda, t)}{\partial t} = a\pi(\lambda, t) - a(\lambda_0 - \lambda)\frac{\partial \pi(\lambda, t)}{\partial \lambda} + (\lambda + b)\pi(\lambda + b, t). \qquad (2.13)$$

However, in such a case λ cannot be a probability magnitude. The difficulty can be overcome by defining λ' by

$$\lambda' = \lambda \quad \text{for } \lambda(t) > 0$$
$$= 0 \quad \text{otherwise.} \tag{2.14}$$

We observe that it is λ' that has probability significance, and in any problem we have to deal only with the moments of λ'.

It is indeed difficult to solve for $\pi(\lambda, t)$ explicitly from (2.12) and (2.13). However, it is possible to obtain the moments of λ'. Defining

$$p(n, t) = \int_{-\infty}^{\infty} \pi(\lambda, t)\lambda' \, d\lambda, \tag{2.15}$$

we obtain

$$\frac{\partial p(n, t)}{\partial t} = -nap(n, t) + na\lambda_0 p(n - 1, t) + \sum_{i=1}^{n} \binom{n}{i} p(n - i + 1, t)(- b)^i \tag{2.16}$$

with the conditions

$$p(0, t) = 1, \quad p(n, 0) = \lambda_0^n. \tag{2.17}$$

The first few moments can be explicitly calculated:

$$p(1, t) = a\lambda_0/(a + b) + b\lambda_0 \exp[- (a + b)t]/(a + b), \tag{2.18}$$

$$p(2, t) = b^2\lambda_0(2\lambda_0 - a - b)[\exp - 2(a + b)t]/2(a + b)^2 +$$
$$b\lambda_0(2a\lambda_0 + b^2) \exp[- (a + b)t]/(a + b)^2 +$$
$$a\lambda_0(2a\lambda_0 + b^2)/2(a + b)^2, \tag{2.19}$$

$$p(3, t) = \{\lambda_0^3 - \lambda_0(a\lambda_0 + b^2)(2a\lambda_0 + b^2)(3b + a)/2(a + b)^3 +$$
$$3b^2\lambda_0(2\lambda_0 - a - 2b)(a\lambda_0 + b^2)/2(a + b)^2 + b^4\lambda_0/2(a + b)^2 +$$
$$b^3 a\lambda_0/3(a + b)^2\} \exp[- 3(a + b)t] +$$
$$\{3b^2\lambda_0(2\lambda_0 - a - 2b)(a\lambda_0 + b^2) \exp[- 2(a + b)t]\}/2(a + b)^3 +$$
$$\{3b\lambda_0(a\lambda_0 + b^2)(2a\lambda_0 + b^2)/2(a + b)^3 -$$
$$b^4\lambda_0/2(a + b)^2\} \exp[- (a + b)t] -$$
$$ab^3\lambda_0/3(a + b)^2 + a\lambda_0(a\lambda_0 + b^2)(2a\lambda_0 + b^2)/2(a + b)^3. \tag{2.20}$$

The moments have a simple form if we proceed to the limit as t tends to infinity. Defining $p(1), p(2), p(3)$ as the limits of $p(1, t), p(2, t),$ and $p(3, t)$, respectively, we find

$$p(1) = a\lambda_0/(a + b),$$

$$p(2) = [p(1)]^2 + b^2 p(1)/2(a + b), \tag{2.21}$$

$$p(3) = [p(1)]^3 + b^2[p(1)]^2/2(a + b)$$
$$- ab^2 p(1)/3(a + b) + b^2 p(2)/(a + b). \tag{2.22}$$

In the special case, when $b = 0$, we find

$$p(n, t) = p(n) = \lambda_0{}^n, \tag{2.23}$$

a result consistent with the facts that the parameter λ is no longer random and that the value of λ at any time should be equal to its initial value λ_0.

2.2. Correlation of Events on the t-Axis

The cumulative response as described by (2.6) and (2.7) is a linear functional of the random variable $dn(\tau)$ representing the number of electron arrivals in the interval $(\tau, \tau + d\tau)$. Although it is difficult to obtain the p.f.f. of $\psi(t)$, it is possible to determine the first few moments and correlations (of the first few orders) of $\psi(t)$. To achieve this, it is first necessary to obtain explicitly the correlations of $dn(\tau)$ corresponding to different values of τ. For any general point process, these correlations are known as product densities (see Ramakrishnan [12, 13]). Defining the product densities of degree 1 and 2 by

$$f_1(t)\, dt = E\{dn(t)\}, \tag{2.24}$$

$$f_2(t_1, t_2)\, dt_1\, dt_2 = E\{dn(t_1)\, dn(t_2)\}, \tag{2.25}$$

we find

$$f_1(t)\, dt = \int_\lambda \pi(\lambda, t)\, d\lambda\, \lambda'\, dt \tag{2.26}$$

$$= \int_0^\infty \lambda \pi(\lambda, t)\, d\lambda\, dt$$

$$= p(1, t)\, dt. \tag{2.26}$$

Thus the mean number of events that have occurred between 0 and t is given by

$$E\{n(t)\} = \int_0^\infty f_1(t)\, dt = a\lambda_0 t/(a + b) -$$

$$b\lambda_0[1 - \exp -(a + b)t]/(a + b)^2. \tag{2.27}$$

To obtain the second-order product density, it is necessary to introduce the function $\pi(\lambda_2, t_2 | \lambda_1, t_1)$, where $\pi(\lambda_2, t_2 | \lambda_1, t_1)\, d\lambda_2$ denotes the probability that $\lambda(t_2)$ has a value between λ_2 and $\lambda_2 + d\lambda_2$, given that $\lambda(t)$ had a value λ_1 at $t = t_1$ and a value λ_0 at $t = 0$:

$$f_2(t_1, t_2) = \int_{\lambda_1} \int_{\lambda_2} \pi(\lambda_1, t_1)\, d\lambda_1\, \lambda_1' \cdot \pi(\lambda_2, t_2 | \lambda_1 - b, t_1)\, d\lambda_2 \lambda_2'$$

$$= \int_{\lambda_1} E\{\lambda'(t_2) | \lambda_1 - b, t_1\} \lambda_1' \pi(\lambda_1, t_1)\, d\lambda_1, \tag{2.28}$$

where $E\{\lambda'(t_2)|\lambda_1 - b, t_1\}$ is the conditional expectation of λ' at t_2 given that λ has a value $\lambda_1 - b$ at t_1. To obtain the conditional expectation, we write the differential equation governing the conditional p.f.f.:

$$\frac{\partial\pi(\lambda_2, t_2|\lambda_1, t_1)}{\partial t_2} = - a(\lambda_0 - \lambda_2)\frac{\partial\pi(\lambda_2, t_2|\lambda_1, t_1)}{\partial\lambda_2} +$$

$$(a - \lambda_2)\pi(\lambda_2, t_2|\lambda_1, t_1) + (b + \lambda_2)\pi(\lambda_2 + b, t_2|\lambda_1, t_1), \quad (2.29)$$

with the initial condition

$$\pi(\lambda_2, t_1|\lambda_1, t_1) = \delta(\lambda_2 - \lambda_1). \quad (2.30)$$

Equation (2.29) holds only if $\lambda_2(t) > 0$. If $\lambda_2(t)$ drops to zero or negative values, its p.f.f. satisfies an equation similar to (2.12). Although the explicit solution of (2.30) is very difficult to achieve, we can obtain the conditional moments of λ', which are precisely the quantities that are necessary for the explicit calculation of the correlation functions. Defining

$$p(n, t_2|\lambda_1, t_1) = E\{[\lambda'(t_2)]^n|\lambda_1, t_1\}, \quad (2.31)$$

we obtain

$$\frac{\partial p(n, t_2|\lambda_1, t_1)}{\partial t_2} = - nap(n_1, t_2|\lambda_1, t_1) + na\lambda_0 p(n - 1, t_2|\lambda_1, t_1) +$$

$$\sum_{i=1}^{n}\binom{n}{i}p(n - i + 1, t_2|\lambda_1, t_1)(- b)^i. \quad (2.32)$$

The first few moments can be explicitly calculated. They are given by

$$p(1, t_2|\lambda_1, t_1) = a\lambda_0[1 - \exp - (a + b)(t_2 - t_1)]/(a + b) +$$
$$\lambda_1 \exp - (a + b)(t_2 - t_1), \quad (2.33)$$

$$p(2, t_2|\lambda_1, t_1) = a\lambda_0(2a\lambda_0 + b^2)/2(a + b)^2 +$$
$$\{\lambda_1{}^2 - (2a\lambda_0 + b^2)[2\lambda_1 - a\lambda_0/(a + b)]/2(a + b)\} \cdot$$
$$\exp[- 2(a + b)(t_2 - t_1)] +$$
$$(2a\lambda_0 + b^2)(\lambda_1 - a\lambda_0/(a + b)) \cdot$$
$$\exp[- (a + b)(t_2 - t_1)]/(a + b). \quad (2.34)$$

From (2.33) and (2.28), we readily obtain

$$f_2(t_1, t_2) = p(2, t_1) \exp[- (a + b)(t_2 - t_1)] +$$
$$p(1, t_1)[a\lambda_0\{1 - \exp - (a + b)(t_2 - t_1)\}/(a + b) -$$
$$b \exp - (a + b)(t_2 - t_1)]. \quad (2.35)$$

In a similar manner we can obtain $f_3(t_1, t_2, t_3)$. An explicit expression for it is given in [11].

An interesting feature that is reflected from (2.35) is the existence of the limit of $f_2(t_1, t_2)$, where both t_1 and t_2 tend to infinity in such a manner that $t_2 - t_1$ remains a constant equal to τ. In fact, we have

$$\lim f_2(t_1, t_2) = [a\lambda_0/(a + b)]^2 - a\lambda_0 b(2a + b) \exp[-(a + b)\tau]/2(a + b)^2$$
$$= [\lim f_1(t_1)]^2 - a\lambda_0 b(2a + b) \exp[-(a + b)\tau]/2(a + b)^2,$$

$$(2.36)$$

demonstrating the persistence of the negative correlation even after infinite time. In Section 2.3 we use these results to obtain the moments and correlations of the cumulative response. It is interesting to note that (2.36) has been conjectured by Rowland [14] on the basis of the experimental findings of Moullin.

2.3. Moments and Correlations of the Cumulative Response

The results of Section 2.2 enable us to obtain the correlation of the response function at two different times after the system has attained equilibrium. Taking the mean value of both sides of (2.7), we find

$$E\{\psi(t)\} = \int_0^t f_1(\tau)\phi(t - \tau) \, d\tau, \qquad (2.37)$$

where $f_1(\tau)$ is given by (2.26). As t tends to infinity, we notice

$$E\{\psi(\infty)\} = \frac{a\lambda_0}{a + b} \int_0^\infty \phi(\tau) \, d\tau - \frac{a\lambda_0}{a + b} \lim \int_0^t e^{-(a+b)(t-\tau)} \phi(\tau) \, d\tau$$

$$= \frac{a\lambda_0}{a + b} \int_0^\infty \phi(\tau) \, d\tau, \qquad (2.38)$$

a result which brings out the experimentally observed fact that the mean value of the cumulative response is reduced from its Poisson value. The second-order correlation of $\psi(t)$ is given by

$$E\{\psi(t_1)\psi(t_2)\} = \int_0^{t_1} \int_0^{t_2} E\{dn(\tau_1) \, dn(\tau_2)\}\phi(t_1 - \tau_1)\phi(t_2 - \tau_2). \quad (2.39)$$

The integrand on the right-hand side can be expressed in terms of the product densities of the pulses if we take into account the degeneracy arising from the overlapping of the intervals $d\tau_1, d\tau_2$. Thus (2.39) can be rewritten as (see Srinivasan and Vasudevan in [15])

$$E\{\psi(t_1)\psi(t_2)\} = \int_0^{\min(t_1,t_2)} f_1(\tau_1)\phi(t_1 - \tau_1)\phi(t_2 - \tau_1) \, d\tau_1 +$$

$$\int_0^{t_1} \int_0^{t_2} \phi(t_1 - \tau_1)\phi(t_2 - \tau_2)f_2(\tau_1, \tau_2) \, d\tau_1 \, d\tau_2. \, (2.40)$$

We are interested in the limiting form of the right-hand side of (2.40) when t_1 and t_2 tend to infinity while $t_2 - t_1$ remains fixed and is equal to τ. It is shown in Appendix of [15] that the Fourier transform of the left-hand side of (2.40), which is a function of the single argument $|t_2 - t_1|$, can be expressed in terms of the Fourier transform of $\phi(t)$ and the limiting form of $f_2(t_1, t_2)$. Thus $r(\omega)$, the power spectrum of the response, is given by*

$$r(\omega) = (2\pi)^{-\frac{1}{2}} a\lambda_0 [|\phi(\omega)|]^2 /(a + b) + 2\pi [a\lambda_0/(a + b)]^2 R|R(\omega)|\phi(\omega)|^2, \quad (2.41)$$

where $R(\omega)$ is the Fourier transform of the limiting form of $f_2(t_1, t_2)$ with respect to the argument $t_2 - t_1$. Equation (2.41) provides a complete description of the second-order property of the shot noise response.

3. BARKHAUSEN NOISE

In the previous section we considered response to shot noise on the basis of a generalized inhomogeneous Poisson process in which the parameter $\lambda(t)$ itself undergoes random changes. A similar situation occurs when a piece of ferro-magnet is magnetized. During the process of magnetization, microscopic Weiss fields start turning and this can be measured as some kind of noise by means of an amplifier. The phenomenon is known as Barkhausen noise, and the average number of pulses per unit time is not necessarily constant (during the process of magnetization) but depends on the instantaneous value of macroscopic magnetization. To incorporate such effects it is necessary to visualize the point process constituted by the train of pulses on the time axis as a general non-Markovian process. The physical quantity of interest is $\psi(t)$, the cumulative response given by

$$\psi(t) = \int_a^t a(\tau)\phi(t - \tau) \, dn(\tau), \quad (3.1)$$

where $dn(\tau)$ represents the number of pulses occurring in the infinitesimal interval $(\tau, \tau + d\tau)$ and $a(\tau)$ is a random amplitude factor. It is assumed that $a(\tau)$ for different τ's are independently and identically distributed. Thus we can develop a good theory of Barkhausen noise by studying an appropriate non-Markovian point process governing the pulse distribution and visualizing the noise as associated response. Two models have been proposed so far; the first one, due to Mazzetti [16], consists in observing that the time interval between any two successive events is governed by a p.f.f. $P(x)$ given by

$$P(x) = v^2 x e^{-vx}, \quad (3.2)$$

where v is positive. This kind of p.f.f. describes a renewal point process and is the simplest example of a non-Markovian process. The other model,

* For a derivation of this formula, see [15].

proposed by Srinivasan and Vasudevan [17], takes into account the variation of the frequency of pulses with the times of occurrence of the previous pulses. In Sections 3.1 and 3.2 we present an account of these two models and derive the power spectrum of the cumulative response.

Before we spell out the details of the specific models, we present the general formulas for the correlation functions. The mean value of $\psi(t)$ is given by

$$\mathbf{E}\{\psi(t)\} = \bar{a} \int_0^t f_1(\tau)\phi(t - \tau)\, d\tau, \tag{3.3}$$

where $\bar{a}$ is the mean value of a and $f_1(\tau)$ is the product density of degree 1 of the point process. The second-order correlation of $\psi(t)$ is given by

$$\mathbf{E}\{\psi(t_1)\psi(t_2)\} = \overline{a^2} \int_0^{\min(t_1, t_2)} \phi(t_1 - \tau)\phi(t_2 - \tau)f_1(\tau)\, d\tau +$$

$$(\bar{a})^2 \int_0^{t_1} \int_0^{t_2} \phi(t_1 - \tau_1)\phi(t_2 - \tau_2)f_2(\tau_1, \tau_2)\, d\tau_1\, d\tau_2, \tag{3.4}$$

where f_2 is the product density of degree 2 of the pulses. If the system reaches equilibrium, the second-order correlation will depend only on the difference $t_1 - t_2 = b$, and its limiting form can be evaluated explicitly:

$$\lim_{\substack{t_1 \to \infty,\, t_2 \to \infty,\\ t_1 - t_2 = b}} \mathbf{E}\{\psi(t_1)\psi(t_2)\} = \rho(b) = I_1 + I_2, \tag{3.5}$$

where

$$I_1 = \overline{a^2} \lim \int_0^{\min(t_1, t_2)} \phi(t_1 - \tau)\phi(t_2 - \tau)f_1(\tau)\, d\tau, \tag{3.6}$$

$$I_2 = \overline{a^2} \lim \int_0^{t_1} \int_0^{t_2} \phi(t_1 - \tau_1)\phi(t_2 - \tau_2)f_2(\tau_1, \tau_2)\, d\tau_1\, d\tau_2. \tag{3.7}$$

3.1. Mazzetti Model

In the model proposed by Mazzetti [16], the time interval distribution x between two successive pulses is governed by (3.2). We assume that a pulse occurs at $t = 0$. Since the process is a stationary renewal point process, the product density of degree 2 is given by

$$f_2(t_1, t_2) = f_1(t_1)f_1(t_2 - t_1). \tag{3.8}$$

Thus the determination of the product density of degree 1 in terms of $P(x)$ will formally complete the solution of the problem. $f_1(t)$ satisfies the equation

$$f_1(t) = P(t) + \int_0^t f_1(\tau)P(t - \tau)\, d\tau. \tag{3.9}$$

Equation (3.9) is obtained by observing that the pulse that occurs between

t and $t + dt$ is either the first pulse or a subsequent pulse. Using the form for $P(t)$ given by (3.2), we can solve for $f_1(t)$ by the Laplace transform technique:

$$f_1(t) = \frac{v}{2}[1 - e^{-2vt}].$$ (3.10)

For the special case

$$\phi(x) = e^{-\alpha x}, \qquad \alpha > 0$$ (3.11)

considered by Mazzetti, we can calculate $\rho(b)$ explicitly. After some calculation, we obtain

$$\rho(b) = \frac{v}{4\alpha}\bar{a}^2 e^{-\alpha|b|} + \bar{a}^2\left[\frac{v^2}{4\alpha^2} + \frac{2v^3}{4\alpha}\frac{e^{-\alpha|b|}}{\alpha^2 - 4v^2} + \frac{v^2 e^{-2v|b|}}{4(4v^2 - \alpha^2)}\right].$$ (3.12)

The shape of the power spectrum obtained by using (3.12) is in agreement with the experimental results (see [16]).

3.2. A Non-Markovian Model

It may be worthwhile to determine whether a more realistic model of the point process governing the pulse formation can explain the Barkhausen effect. If λ_0 is the average rate of occurrence of pulses initially at time $t = 0$, it is eminently reasonable to assume that the rate $\lambda(t)$ at any subsequent time t is given by

$$\lambda(t) = \lambda_0 + b\sum_i \exp - a(t - t_i), \qquad b, a > 0,$$ (3.13)

where the t_i's are the times of occurrence of pulses up to t. Thus $\lambda(t)$ satisfies the stochastic equation

$$\frac{d\lambda}{dt} = -a(\lambda_0 - \lambda) + b\frac{dn(t)}{dt},$$ (3.14)

where $n(t)$ represents the number of pulses that have occurred in $(0, t)$. The cumulative response $\psi(t)$ is again given by (3.1), where $dn(t)$ represents the number of pulses in $(t, t + dt)$, corresponding to a generalized inhomogeneous Poisson process with $\lambda(t)$ governed by (3.13).

Equation (3.13) is very similar to (2.6), and we can follow the same method as in the previous section to arrive at the product densities of the point process. Of course, in the present case, simplicity arises inasmuch as the parameter $\lambda(t)$ never drops to zero. After some calculation, we obtain

$$f_1(t) = [a\lambda_0 - b\lambda_0 \exp - (a - b)t]/(a - b),$$ (3.15)

$$f_2(t_1, t_2) = P(2, t) \exp - (a - b)(t_2 - t_1) +$$
$$f_1(t_1)\{b[\exp - (a - b)(t_2 - t_1)] +$$
$$a\lambda_0(1 - \exp[-(a - b)(t_2 - t_1)])/(a - b)\},$$ (3.16)

where

$$P(2, t_1) = \frac{b^2\lambda_0}{2(a - b)^2}(2b - a + 2\lambda_0)\exp[-2(a - b)t] -$$

$$\frac{b\lambda_0}{(a - b)^2}(2a\lambda_0 + b^2)\exp[-(a - b)t] + \frac{(2a\lambda_0 + b^2)a\lambda_0}{2(a - b)^2}. \quad (3.17)$$

The limiting form of $f_2(t_1, t_2)$ when t_1 and t_2 tend to infinity with $t_2 - t_1 = \tau$ held finite, given by

$$\lim f_2(t_1, t_2) = \left(\frac{a\lambda_0}{a - b}\right)^2 + \frac{ab\lambda_0}{2(a - b)^2}(2a - b)\exp[-(a - b)\tau] \quad (3.18)$$

bears a great similarity to the limiting form obtained in Section 3.2. The power spectrum $\rho(\omega)$ of the cumulative response can be calculated:

$$\rho(\omega) = \frac{a\lambda_0}{a - b}\frac{\overline{a^2}}{2\pi(\alpha^2 + \omega^2)} + \bar{a}^2\left[\frac{2\delta(\omega)}{\alpha^2 + \omega^2}\left(\frac{a\lambda_0}{a - b}\right)^2 + \right.$$

$$\left. \frac{1}{\pi}\frac{ab\lambda_0}{(a - b)}\frac{(2a - b)}{(a - b)^2 + \omega^2}\right]. \quad (3.19)$$

For purposes of illustration we have taken the form $\phi(x)$ given by (3.11). It is interesting to note that the shape of the power spectrum is the same as that given by Mazzetti. However, there is one important feature that makes this model distinct from a renewal process; it is due to the difference in the sign of the coefficient of the exponential term in the limiting form of the correlation function.

4. FLUCTUATION OF PHOTOELECTRONS

So far, we have considered certain response phenomena in the realm of Classical Physics, the response being essentially due to a certain non-Markovian point process. In this section we deal with the photoelectric counts actuated by an incident radiation field. When a radiation field is incident on a fast photodetector, electrons are emitted and the number of electrons emitted in a certain time interval $(0, t)$ is a measure of the intensity of the incident beam. It has been successfully demonstrated by Hanbury-Brown and Twiss [18, 19] that the fluctuations in the photodetected incoherent light is made up of two distinct parts: shot noise (Poisson distribution), and wave interaction or excess photon noise. Thus a Poisson distribution for the number of photocounts in any time interval $(0, T)$ neglects the wave interaction or quantum mechanical nature of the photons (see, for example, Mandel [20] and Mandel, Sudarshan, and Wolf [21]). If we take into account the wave interaction, we are naturally led to a non-Markovian situation.

The problem may be posed in two ways. For a given radiation field (or light beam), what is the probability frequency function of the photoelectrons that are emitted and observed by the detector? Or we can solve the inverse problem. For an observed distribution of photocounts, what are the statistical properties of the incident beam? In fact, the solution to the latter problem is of great interest in Optics. We shall confine our attention to the first question and present some of the recent results obtained by Srinivasan and Vasudevan [22]. This will throw light on the inverse problem, inasmuch as any deviation from the theoretical distribution for the counts corresponding to the Gaussian radiation field can be attributed to the existence of sources emitting a different type of radiation field.

4.1. Photoelectric Counts

From physical considerations it is clear that the probability that an electron is emitted by the detector in the time interval $(t, t + dt)$ is $\alpha I(t)\, dt$, where α is the sensitivity of the detector (taken to be a constant) and $I(t)$, the intensity of the radiation falling on the detector, is given by

$$I(t) = V^*(t)V(t), \tag{4.1}$$

$V(t)$ being the usual analytic signal corresponding to the radiation field. The average number of counts in the time interval $(0, T)$ is given by

$$\bar{n} = \alpha \int_0^T I(t)\, dt. \tag{4.2}$$

If I is a deterministic function of time, the probability of obtaining n counts in $(0, T)$ obeys the Poisson law

$$P(n, T) = \exp\left[- \alpha \int_0^T I(t)\, dt \right]\left[\alpha \int_0^T I(t)\, dt \right]^n \bigg/ n! \tag{4.3}$$

This, of course, corresponds to the classical picture, the wave interaction being completely neglected. A better approximation consists in allowing I to be a random parameter (independent of t) governed by the p.f.f. $P(I)$. In such a case, $P(n, T)$ is given by

$$P(n, T) = \int P(I) \exp[- \alpha IT](\alpha IT)^n / n!\, dI. \tag{4.4}$$

If $P(I)$ is chosen to be an exponential distribution (this corresponds to an uncorrelated Gaussian signal) of the form

$$P(I) = \frac{1}{I_0} \exp\left[- \frac{I}{I_0} \right], \tag{4.5}$$

we obtain

$$P(n, T) = (1 + \bar{n})^{-1}(1 + \bar{n}^{-1})^{-n}, \tag{4.6}$$

where

$$\bar{n} = \alpha I_0 T. \tag{4.7}$$

The preceding result was first obtained by Wolf and Mehta [23]. Although the Bose-Einstein distribution describes the spectrum of the photons, the absence of correlation of the intensity for different times indicates that the result above cannot be applied to the thermal radiation field in which the correlation is inherently built-in. If we take $I(t)$ to be a correlated random process, (4.4) is no longer true. We can circumvent the difficulty by resorting to the product density technique approach, as in the case of shot noise and Barkhausen noise. If $f_1(t_1)$, $f_2(t_1, t_2)$, $f_3(t_1, t_2, t_3)$, ... are the product densities of events (the events corresponding to the ejection of photoelectrons), then it is easy to see that

$$f_1(t_1) = \alpha E\{I(t_1)\},$$
$$f_2(t_1, t_2) = \alpha^2 E\{I(t_1)I(t_2)\},$$
$$f_3(t_1, t_2, t_2) = \alpha^3 E\{I(t_1)I(t_2)I(t_3)\}, \tag{4.8}$$

where the expectation is to be evaluated with the help of the joint probability frequency function of the analytic signals at different times. At first sight, it might appear that the problem is somewhat simpler than the space-charge-limited shot noise. However, the complexity lies in the continuous random nature of $I(t)$.

The mean square number of counts in the interval $(0, T)$ is given by

$$\overline{n^2} = \int_0^T f_1(t)\, dt + \int_0^T \int_0^T f_2(t_1, t_2)\, dt_1\, dt_2. \tag{4.9}$$

To evaluate the right-hand side of (4.2), we assume that the Fourier components of $V(t)$ are distributed according to a Gaussian law so that $\pi(V, t)$ is given by

$$\pi(V, t) = A \exp\left[-\int V^*\left(t + \frac{\tau}{2}\right)V\left(t - \frac{\tau}{2}\right) d\tau \right], \tag{4.10}$$

which, in turn, yields the result

$$E\{I(t_1)I(t_2)\} = E\{I(t_1)\}E\{I(t_2)\} + |\Gamma(t_1 - t_2)|^2, \tag{4.11}$$

where Γ is the coherence function defined by

$$\Gamma(t_1 - t_2) = E\{V^*(t_1)V(t_2)\}. \tag{4.12}$$

If we substitute (4.11) in (4.9), we find that there is a deviation from the simple Poisson law (shot noise) due to the second term on the right-hand side of (4.11). The coherence term arises essentially from the wave inter-interference characteristic of the incident beam. The result above can be readily extended to the higher moments of the count by calculating higher-order correlations of the intensity. In the case of the experiments of Hanbury-Brown and Twiss, the incident intensity is the input of a linear filter, and the power spectrum of the output is measured. We can use the methods outlined in Section 2 to obtain the power spectrum.

4.2. Probability Distribution of the Counts

We can obtain the general moment of the number of counts by introducing the product density generating functional (see Kuznestov *et al.* [24]) by

$$L(u) = E\{\exp \int u(t)I(t) \, dt\}$$

$$= 1 + \sum_{m=1}^{\infty} \frac{1}{m!} \int \int \dots \int f_m(t_1, t_2, \dots, t_m) u(t_1) u(t_2) \dots \cdot$$

$$u(t_m) \, dt_1 \, dt_2 \dots dt_m$$

$$= \exp \sum_{1}^{\infty} \frac{1}{m!} \int \int \dots \int g_m(t_1, t_2, \dots, t_m) u(t_1) u(t_2) \dots \cdot$$

$$u(t_m) \, dt_1 \, dt_2 \dots dt_m, \quad (4.13)$$

where the g_i are the actual correlation functions which are related to the product densities f_i by

$$f_1(t_1) = g_1(t_1),$$

$$f_2(t_1, t_2) = g_1(t_1)g_1(t_2) + g_2(t_1, t_2),$$

$$f_3(t_1, t_2, t_3) = g_1(t_1)g_1(t_2)g_1(t_3) + 3\{g(t_1)g_2(t_2, t_3)\}_{\text{sym}} + g_3(t_1, t_2, t_3). \quad (4.14)$$

The actual correlation functions play an important role inasmuch as they provide a direct measure of the deviation from the Poisson law. It is interesting to note that $g_i(t_1, t_2, \dots, t_i)$ are identical with the coherence functions $\Gamma^i(t_1, t_2, \dots, t_i)$ defined by

$$\Gamma^i(t_1, t_2, \dots, t_i) = E\{V^*(t_1)V^*(t_2) \dots V^*(t_i)V(t_1)V(t_2) \dots V(t_i)\}. \quad (4.15)$$

Thus it is clear from (4.13) that, once we are in possession of all the moments of the number distribution, we can readily obtain the magnitude and phase of all the coherence functions. In particular, if, in a certain type of beam, coherence functions of all orders less than or equal to l determine all the higher-order coherence functions, then a knowledge of the first $(l + 1)$ moments of the number distribution will be sufficient to determine all the coherence functions, including their phases.

To obtain $P(n, T)$ we notice that the probability generating function defined by

$$h(z, T) = \sum_n P(n, T)z^n \quad (4.16)$$

can be obtained from $L(u)$:

$$L(z - 1) = h(z, T). \tag{4.17}$$

Thus we can use (4.13) to obtain L explicitly if we make some reasonable assumptions for the coherence functions. For thermal light, the intensities are distributed according to the Gaussian law:

$$\Gamma^m(t_1, t_2 \ldots, t_m) = (m - 1)!\Gamma^2(t_1 - t_2)\Gamma^2(t_2 - t_3)\ldots\Gamma^2(t_m - t_1). \tag{4.18}$$

If, in addition, we assume that $\Gamma^2(t_1 - t_2) = \alpha\bar{I}$, we obtain

$$L[u] = \left[1 - \alpha I \int_0^T u(t)\, dt\right]^{-1} \tag{4.19}$$

from which we find

$$h(z, T) = [1 + \alpha\bar{I}T - \alpha\bar{I}zT]^{-1}, \tag{4.20}$$

a result which identifies $P(n, T)$ with the Bose–Einstein distribution (4.6).

For nonthermal beams, (4.17) expresses the probability generating function in terms of the coherence functions. This is a significant result because it enables us to compute, at least numerically, the probability frequency function of the photo counts for any arbitrary radiation field.

5. RESPONSE OF CONTINUOUS STRUCTURES

The study of the structural response of mechanical systems subject to random loadings has received considerable importance of late in view of some recent developments in jet and rocket propulsion. The pressure fields generated by these devices fluctuate in a random manner and very often contain a wide spectrum of frequencies resulting in severe vibration of the system. Thus a general study of response phenomena of continuous structures, taking into account their nonlinear characteristics, will throw light on many of modern structural problems. In fact, almost all the statistical problems of continuum mechanics can be formulated as response phenomena,* and in this section we confine our attention to the class of problems that can be discussed from the viewpoint mentioned in Section 1. We discuss other types of problems of continuum mechanics in the next chapter.

Since a continuous structure is a mechanical system with an infinite number of degrees of freedom, the most natural approach to such a problem is the classical method of the normal modes which are obtained from the equation for undamped free motion. The continuous structure that we consider in this section will be either one-dimensional or two-dimensional in nature. Pioneering studies in the statistical properties of such a structure can be

* It is interesting to note that even the general problem of turbulent motion of fluids can be construed as some kind of a response due to random functionals (see, for example, Srinivasan [25]).

traced to Van Lear and Uhlenbeck [26], who used the impulse-response method and studied both stationary and nonstationary solutions and their characteristics. There have been further attempts in this direction, particularly by Lyon [27], Dyer [28], Bogdanoff and Goldberg [29], Thompson and Barton [30], Eringen [31] and Powell [32], who have confined their attention to stationary solutions. More recently, the nonstationary aspects of the response problem have been analysed by Lin [33] and Roberts [34] (see also [35]). In Section 5.1 we present an account of the response of systems subjected to random loading. In Section 5.2 we deal with the problem of fatigue failure.

5.1. Random Loading

As we have observed in the introductory remarks above, the earliest attempts at determining the response of continuous structure to random force fields are those of Van Lear and Uhlenbeck [26], who dealt with the Brownian motion of strings and rods. Let us first consider the case of the homogeneous string. The displacement u satisfies the equation

$$\rho\frac{\partial^2 u}{\partial t^2} + f\frac{\partial u}{\partial t} = \frac{\partial}{\partial x}\left(\tau\frac{\partial u}{\partial x}\right) + F(x, t). \tag{5.1}$$

The string, assumed to be of length L and bound elastically at its ends, is surrounded by a gas and is under tension $\tau(x)$. The symbol f stands for the frictional coefficient, and $F(x, t)$ is the fluctuating force to which the string is subjected. Equation (5.1) may be written in the form

$$\frac{\partial^2 u}{\partial t^2} + \beta\frac{\partial u}{\partial t} = p'\frac{\partial u}{\partial x} + p\frac{\partial^2 u}{\partial x^2} + A(x, t), \tag{5.2}$$

where

$$\beta = \frac{f}{\rho}, \quad p = \frac{\tau}{\rho}, \quad \text{and} \quad A = \frac{F}{\rho}. \tag{5.3}$$

The boundary conditions are given by

$$h_0 u(0, t) - \left(\frac{\partial u}{\partial x}\right)_{x=0} = 0,$$

$$h_L u(L, t) - \left(\frac{\partial u}{\partial x}\right)_{x=L} = 0, \tag{5.4}$$

where h_0 and h_L are the ratios of the elastic constants of binding to tension at the two ends.

The random function $A(x, t)$ satisfies the usual conditions:

$$\mathbf{E}\{A(x, t)\} = 0,$$

$$\mathbf{E}\{A(x_1, t_1)A(x_2, t_2)\} = \phi(x_1 - x_2, t_1 - t_2), \tag{5.5}$$

where ϕ is an even function in both x and t, having a sharp maximum at $(0, 0)$.

Van Lea and Uhlenbeck assumed that the random force is not correlated in space, and they solved the stochastic equations by the variable separable method. They obtained the formal solution in terms of the eigenvalues of the homogeneous equation corresponding to (5.3) and estimated the mean square value of the displacement. An exactly similar procedure was adopted by them for studying the response of an elastic rod subject to a fluctuating force.

Lin [33] extended the results to a more general case by studying the differential equation

$$M\ddot{w} + C\dot{w} + Q(w) = p(\mathbf{r}, t), \tag{5.6}$$

where Q is a linear (either one-dimensional or two-dimensional) differential operator in the spatial variables. We have, for example,

$$Q = -T\left(\frac{\partial^2}{\partial x^2}\right) \quad \text{for a taut string}, \tag{5.7}$$

$$Q = D\left(\frac{\partial^4}{\partial x^4} + 2\frac{\partial^4}{\partial x^2 \partial y^2} + \frac{\partial^4}{\partial y^4}\right) \quad \text{for a flat plate},$$

where T is the tensile force in the string and $D\,[\,= Eh^3/12(1 - v^2)]$ is the bending rigidity of the plate. The solution of (5.6) in terms of the Green's function $g(\mathbf{r}, t; \mathbf{r}', t')$ is given by

$$w(\mathbf{r}, t) = \int_0^t \int_R g(\mathbf{r}, t; \mathbf{r}', t')p(\mathbf{r}', t')\,dt'\,d\mathbf{r}' + w_c(\mathbf{r}, t), \tag{5.8}$$

where $w_c(\mathbf{r}, t)$ is the solution to the homogeneous equation obtained by setting $p(\mathbf{r}, t)$ to zero. Thus $w_c(\mathbf{r}, t)$ depends on the boundary and initial conditions.

Following Van Lear and Uhlenbeck [26], if we assume that the solution is expressible in terms of the normal modes, we can calculate the Green's function explicitly. Since the method of calculation involves only standard techniques, we give the final result arrived at by Lin:

$$g(r, t; r', t') = \sum_{n=1}^{\infty} \frac{f_n(\mathbf{r})f_n(\mathbf{r}')}{M_n \omega_{nd}} e^{-\zeta_n \omega_n(t-t')} \sin \omega_{nd}(t - t')H(t - t'), \tag{5.9}$$

where

$$M_n = \int_R M f_n^{\,2}(\mathbf{r})\,d\mathbf{r} \quad \text{and} \quad C_n = \int_R C f_n^{\,2}(\mathbf{r})\,d\mathbf{r} \tag{5.10}$$

are, respectively, the generalized mass and damping in the nth mode. The quantity ω_n is the frequency of the undamped motion, and ζ_n and ω_{nd} are given by

$$\zeta_n = \frac{C_n}{2M_n\omega_n}, \qquad \omega_{nd} = (1 - \zeta_n^2)^{-\frac{1}{2}}\omega_n, \qquad (5.11)$$

and $H(t - t')$ is the Heaviside unit function. Equations (5.8) and (5.9) constitute the formal solution of the problem. The mean value of $w(\mathbf{r}, t)$ is given by

$$\mathbf{E}\{w(\mathbf{r}, t)\} = \int_0^t \int_R g(\mathbf{r}, t; \mathbf{r}', t')\mathbf{E}\{p(\mathbf{r}', t')\} \, dr' \, dt' + w_c(\mathbf{r}, t), \qquad (5.12)$$

so that we have the result

$$\mathbf{E}\{w(\mathbf{r}, t)\} = w_c(\mathbf{r}, t) \qquad (5.13)$$

since we can set $\mathbf{E}\{p(\mathbf{r}, t)\} = 0$ without loss of generality. If we assume also $w_c(\mathbf{r}, t) = 0$, the second-order correlation of $w(\mathbf{r}, t)$ is given by

$$\mathbf{E}\{w(\mathbf{r}, t_1)w(\mathbf{r}_2, t_2)\} = \sum_{n=1}^{\infty} \sum_{m=1}^{\infty} \mathbf{E}\{w_n(\mathbf{r}_1, t_1)w_m(\mathbf{r}_2, t_2)\}, \qquad (5.14)$$

$$\mathbf{E}\{w_n(\mathbf{r}_1, t_1 w)_m(\mathbf{r}_2, t_2)\} =$$

$$\int_0^{t_1} dt_1' \int_0^{t_2} dt_2' \int_R dr_1' \int_R dr_2' \mathbf{E}\{p(\mathbf{r}_1', t_1')p(\mathbf{r}_2', t_2')\} \cdot$$

$$f_n(\mathbf{r}_1)f_n(\mathbf{r}_1')e^{-\zeta_n\omega_n(t_1-t_1')} \sin \omega_{nd}(t_1 - t_1') \cdot$$

$$f_m(\mathbf{r}_2)f_m(\mathbf{r}_2')e^{-\zeta_m\omega_m(t_2-t_2')} \sin \omega_{nd}(t_2 - t_2'). \qquad (5.15)$$

If the random forcing field is stationary with respect to t and homogeneous with respect to $\mathbf{r}$, (5.14) can be written in the form

$$\mathbf{E}\{w_n(\mathbf{r}_1, t_1)w_m(\mathbf{r}_2, t_2)\} =$$

$$\int_{-\infty}^{+\infty} e^{i\omega(t_1-t_2)} I_n(\omega)\frac{f_n(\mathbf{r}_1)f_m(\mathbf{r}_2)}{Z_nZ_m^*}\chi_n(t_1, \omega)\chi_m^*(t_2, \omega) \, d\omega, \qquad (5.16)$$

$$\chi_n(t, \omega) = 1 - e^{-\zeta_n\omega_nt-i\omega t}\left[\cos \omega_{nd}t + \frac{\zeta_n\omega_n + i\omega}{\omega_{nd}} \sin \omega_{nd}t\right]. \qquad (5.17)$$

In (5.16) the asterisk denotes the complex conjugate, and

$$Z_n = M_n(\omega_n^2 - \omega^2 + 2i\zeta_n\omega_n\omega) \qquad (5.18)$$

is the impedance in the nth mode. The function $I_{nm}(\omega)$ is given by

$$I_{nm}(\omega) = \int_R dr_1' \int_R dr_2'\Phi(r_1' - r_2', \omega),$$

$$\Phi(r_1' - r_2', t_1' - t_2') = \int_{-\infty}^{+\infty} \Phi(r_1' - r_2', \omega)e^{i\omega(t_1'-t_2')} \, d\omega, \qquad (5.19)$$

where ϕ, defined as the expectation value of the product $p(r_1', t_1')p(r_2', t_2')$, is a function of $r_1' - r_2'$ and $t_1' - t_2'$ only in view of the homogeneous and stationary nature of the field.

If we specialize further by taking $\mathbf{r}_1 = \mathbf{r}_2 = \mathbf{r}$, and $t_1 = t_2 = t$ and allow t to tend to infinity, we obtain

$$\mathbf{E}\{[w(\mathbf{r})]^2\} = \sum_{n=1}^{\infty} \sum_{m=1}^{\infty} f_n(\mathbf{r})f_m(\mathbf{r}) \int_{-\infty}^{+\infty} \frac{I_{nm}}{Z_n Z_m^*} \, d\omega. \tag{5.20}$$

The result above was derived by Powell [32], and the more general relation (5.1) is due to Lin [33]. On the basis of (5.20), Lin discussed the response of a single-degree-of-freedom system by considering a simple mass-spring-dashpot system.

Lin [36] also considered the response of such a system to nonstationary shot noise. He defined the nonstationary shot noise process as a weighted Poisson process. More specifically, the process is defined by the random variable $F(t)$,

$$F(t) = x(t) \, dn(t), \tag{5.21}$$

where $\{x(t)\}$ for different values of t denotes a set of random variables independently and identically distributed such that $\mathbf{E}\{x(t)\} = 0$. $n(t)$ denotes the number of pulses in $(0, t)$ obeying a Poisson law. On the basis of equation (5.22), it is quite easy to calculate the characteristics of the linear response to the nonstationary shot noise. This was done by Srinivasan and Kumaraswamy [37], who obtained the following expression for the characteristic functional $C[\theta]$:

$$C[\theta] = \exp - \int_0^T \lambda \left[1 - \int_0^\infty \cos[x(t)\theta(t)]p(x, t) \, dx \right] dt, \tag{5.22}$$

where $p(x, t)$ is the p.f.f. governing $x(t)$.

From the expression above, these authors concluded that, while all the odd-order correlations of $F(t)$ vanish, the even-order correlations are not capable of being expressed as the sum of all possible products of second-order correlations. For instance, the fourth-order correlation is given by

$\mathbf{E}\{F(t_1)F(t_2)F(t_3)F(t_4)\} =$

$$\mathbf{E}\{F(t_1)F(t_2)\}\mathbf{E}\{F(t_3)F(t_4)\} +$$
$$\mathbf{E}\{F(t_1)F(t_3)\}\mathbf{E}\{F(t_2)F(t_4)\} +$$
$$\mathbf{E}\{F(t_1)F(t_4)\}\mathbf{E}\{F(t_2)F(t_3)\} +$$
$$\lambda_i \int_0^\infty [x(t_1)]^4 p(x, t_1) \, dx \, \delta(t_2 - t_1) \, \delta(t_3 - t_1) \, \delta(t_4 - t_1). \tag{5.23}$$

Thus we can conclude that the nonstationary shot noise introduced by Lin [36] (as a possible model for a Gaussian process) cannot be made a Gaussian process.

We finally observe that our discussion in this section pertains only to the response of a linear structure to random excitation. However, in practice, we do encounter nonlinear structures (see, for example, [38, 39]), and in such a case we may have to use the approximate methods introduced in Section 4 of Chapter 3. The response phenomena of nonlinear structures was studied by Smith [40] and Caughey [41], and we discuss this problem again in Chapter 6 when we study stochastic problems in continuum mechanics.

5.2. Fatigue Failure

In this section we formulate the fatigue problem of materials as a response phenomenon and show how the salient features can be deduced from the viewpoint of stochastic processes. The term "fatigue" means injury or damage leading to cracking as the result of repeated stress. It infers a process of localized progressive structural change occurring in material subjected to fluctuating stress which generally results in lowering the resistance or reliability of the material for subsequent stresses. Seldom does a single application of a static load cause material failure. On the other hand, repeated applications of a load of lesser magnitude, though considered to be a safe load from static design considerations, may ultimately cause sudden and catastrophic failure. Such failures are termed fatigue failures.

A method of treating the fatigue of a metal under a spectrum of loading is to make use of the classical linear damage rule first proposed by Palmgreen and later by Langer and Miner (for example, see [42]). They assume that at any stage of the loading history of the material, a percentage of life used up is proportional to the cycle ratio at that loading condition. Thus, if a stress range is applied for n_1 cycles at a condition where failure would occur if N_1 cycles were applied, the percentage of life used up is n_1/N_1. It is also assumed that the fractional damage at this stress level can be added to the corresponding fractional damages of other stress levels to obtain the total damage experienced by the material. Thus, if the specimen experiences n_i cycles of stress amplitudes s_i, $i = 1, 2, \ldots$, the total cumulative damage fraction is taken to be given by

$$D = \sum \frac{n_i}{N_i},$$ (5.24)

and failure is said to occur when D reaches unity. Parzen [43] studied this problem by the use of renewal theory and deduced a criterion for the failure.

Srinivasan and Kumaraswamy [44] (see also [45]) also analyzed this problem by writing the cumulative damage $\mathscr{D}$ at the end of the loading history as

$$\mathscr{D} = \int_0^t D[s(\tau)]\, dN(\tau), \qquad (5.25)$$

where $N(t)$ represents the number of random points $t_1, t_2, \ldots, t_i, \ldots$ in the interval $(0, t)$ of the time axis when the stresses have been applied and $D[s(\tau)]$ is the damage experienced by the material from the stress $S(\tau)$ applied at time τ. The mean value of $\mathscr{D}$ exceeding the strength of the material, together with some knowledge of fluctuation about the mean, may be taken to be a reasonable criterion for the failure of the material.

The notion of cumulative damage is not a realistic criterion because the very concept of fractional damage is not precise. Recognizing this aspect, Valluri [46] proposed a theory of metal fatigue based on the assumption that the propagation of a "dominant" crack is primarily responsible for the ultimate catastrophic failure of the specimen. The abstract and perhaps ill-defined notion of damage is replaced by the precise and realistic one of "crack length." In the approach of Valluri the fatigue process is considered to take place via the formation, propagation, and eventual catastrophic failure of a dominant crack. The last stage of the process occurs when crack length l attains a critical value for the applied stress s, l being given by the Griffith criterion $l = l_{cr}(s) = c/s^2$, where c is some parameter. If we assume that the material has a crack of length l_0 (sufficient to propagate), the fatigue process is then a study of the growth of a single crack from a length l_0 to a length sufficient for the final failure. Srinivasan and Kumaraswamy [44] analyzed this problem by proposing three different models. In the next few paragraphs we summarize the salient features of one of the models based on the critical stress.

We first observe that Griffith's criterion can be written in the form

$$s = s_u \sqrt{\frac{l_0}{l}}, \qquad (5.26)$$

where s_v is defined to be that stress for which the critical crack length is l_0. The law of propagation of the crack is taken as a function of the applied stress, say $F(s)$. For a simple load application at the stress level s, the crack length grows from l_1 to l_2, given by the relation

$$\frac{2}{l_1^{\frac{1}{2}}} - \frac{2}{l_2^{\frac{1}{2}}} = F(s), \qquad (5.27)$$

so that a sequence of stresses $s_1, s_2, \ldots, s_n$ extends the crack from l_0 to l such that

$$\frac{2}{l_0^{\frac{1}{2}}} - \frac{2}{l^{\frac{1}{2}}} = \sum_{i=1}^n f(s_i). \qquad (5.28)$$

We shall consider the case when the material is subjected to random stresses $s_1, s_2, \ldots, s_n$ at random instants of time, $t_1, t_2, \ldots, t_n$, on the time axis.* Thus (5.28) can be generalized and written as the stochastic equation

$$\frac{(l_0)^{\frac{1}{2}}}{l} = 1 - \frac{l_0^{\frac{1}{2}}}{2} \int_0^t F(s(\tau)) \, dN(\tau). \tag{5.29}$$

From the Griffith criterion, the stress S needed to cause failure at time t is given by

$$S(t) = s_u \left[1 - \frac{l_0^{\frac{1}{2}}}{2} \int_0^t F(s(\tau)) \, dN(\tau) \right]. \tag{5.30}$$

The mean value of this stress is given by the formal expression

$$\mathbf{E}\{S\} = s_u \left[1 - \frac{l_0^{\frac{1}{2}}}{2} \int_0^t \mathbf{E}\{F(s(\tau))\} f_1(\tau) \, d\tau \right], \tag{5.31}$$

where $f(\tau)$ is the product density of degree 1 of the instants of application of the stresses. However, it should be noted that in the calculation of $\mathbf{E}\{F(s(\tau))\}$, the upper limit for s is again determined by the relation

$$S(\tau) = s_u \left[1 - \frac{l_0^{\frac{1}{2}}}{2} \int_0^\tau F(s(\tau')) \, dN(\tau') \right]. \tag{5.32}$$

Thus the problem posed in this manner does not appear to lend itself for analytical treatment on account of the inherent nonlinear nature of (5.30). Lardner [47] has, for a certain model of propagation and failure criterion for the risk function, reduced the exact equation to an algebraic equation which he has solved by successive approximations. However, it may be worthwhile to tackle the problem directly by more powerful techniques like the quasi-linearization method after converting the integral relation (5.29) into differential form.

* In practice, the stresses experienced by the material at different times are random, and the instants of applications of such stresses are themselves random in nature (see, for example, Lardner [47]). One of us (S.K.S.) is grateful to Professor Valluri for clarifying this point.

REFERENCES

1. N. Campbell, *Proc. Cambridge Philos. Soc.*, **15**(1909), 117.
2. S. O. Rice, *Bell. Syst. Tech. J.*, **23**(1944), 282.
3. J. E. Moyal, *J. Roy. Statist. Soc.*, **B11**(1949), 150.
4. E. N. Rowland, *Proc. Cambridge Philos. Soc.*, **32**(1936), 580.
5. A. W. Hull and N. H. Williams, *Phys. Rev.*, **25**(1925), 147.
6. J. B. Johnson, *Phys. Rev.*, **26**(1925), 71.

7. E. B. Moullin, *Proc. Roy. Soc. (London)*, **A147**(1934), 100.
8. E. N. Rowland, *Proc. Cambridge Philos. Soc.*, **33**(1937), 344.
9. S. K. Srinivasan, On a Class of non-Markovian Processes, IIT Preprint, August 1962.
10. M. S. Bartlett, *J. Roy. Statist. Soc.*, **B25**(1963), 264.
11. S. K. Srinivasan, *Nuovo Cimento*, **38**(1965), 979.
12. Alladi Ramakrishnan, *Proc. Cambridge Philos. Soc.*, **46**(1950), 595.
13. Alladi Ramakrishnan, *Proc. Cambridge Philos. Soc.*, **49**(1953), 473.
14. E. N. Rowland, *Proc. Cambridge Philos. Soc.*, **34**(1938), 329.
15. S. K. Srinivasan, in *Proceedings of Symposia in Theoretical Physics*, Vol. 3, Alladi Ramakrishnan, ed., Plenum Press, New York, 1966.
16. P. Mazzetti, *Nuovo Cimento*, **25**(1962), 1322; **31**(1964), 88.
17. S. K. Srinivasan and R. Vasudevan, *Nuovo Cimento*, **41**(1966), 101.
18. R. Hanbury-Brown and R. Q. Twiss, *Philos. Mag.*, **45**(1954), 663.
19. R. Hanbury-Brown and R. Q. Twiss, *Proc. Roy. Soc. (London)*, **A242**(1957), 300; **A243** (1957), 291.
20. L. Mandel, *Proc. Phys. Soc. (London)*, **81**(1963), 1104.
21. L. Mandel, E. C. G. Sudarshan, and E. Wolf, *Proc. Phys. Soc.*, **84**(1964), 435.
22. S. K. Srinivasan and R. Vasudevan, *Nouvo Cimento*, **47**(1967), 185.
23. E. Wolf and C. L. Mehta, *Phys. Rev. Lett.*, **13**(1964), 705.
24. P. I. Kuznestov, R. L. Stratonovich, and V. I. Tikhnov, *Non-linear Transformation of Stochastic Processes*, Pergamon Press, London, Chap. I, Sec. 6.
25. S. K. Srinivasan, *Z. fur Physik*, **197**(1966), 435; **205**(1967), 221.
26. G. A. Van Lear, Jr., and G. E. Uhlenbeck, *Phys. Rev.*, **38**(1931), 1583.
27. R. H. Lyon, *J. Acoust. Soc. Amer.*, **28**(1956), 391.
28. I. Dyer, *J. Acoust. Soc. Amer.*, **31**(1959), 922.
29. J. L. Bogdanoff and J. E. Goldberg, *J. Aerospace Sci.*, **27**(1960), 371.
30. W. T. Thompson and M. V. Barton, *J. Appl. Mech.*, **24**, *Trans. ASME*, **79**(1957), 248.
31. A. C. Eringen, *J. Appl. Mech.*, **24**, *Trans. ASME*, **79**(1957), 46.
32. A. Powell, *J. Acoust. Soc. Amer.*, **30**(1958), 1130.
33. Y. K. Lin, *J. Acoust. Soc. Amer.*, **35**(1963), 222; **38**(1965), 453.
34. J. B. Roberts, *J. Sound Vibr.*, **2**(1965), 336, 375.
35. S. K. Srinivasan, R. Subramanian, and S. Kumaraswamy, *J. Sound. Vibr.*, **6**(1967), 169.
36. Y. K. Lin, *J. Acoust. Soc. Amer.*, **36**(1964), 82.
37. S. K. Srinivasan and S. Kumaraswamy, *J. Appl. Mech., Trans. ASME*, **37** (1970) 543.
38. R. H. Lyon, M. Heckl, and C. B. Hazelgrove, *J. Acoust. Soc. Amer.*, **33**(1961), 1404.
39. P. W. Smith, Jr., C. I. Malme, and C. M. Gogos, *J. Acoust. Soc. Amer.*, **33** (1961), 1476.
40. P. W. Smith, Jr., *J. Acoust. Soc. Amer.*, **34**(1962), 827.
41. T. K. Caughey, *J. Acoust. Soc. Amer.*, **35**(1963), 1683.
42. S. H. Crandall and D. Mark, *Random Vibration in Mechanical Systems*, Academic Press, New York, 1963.
43. E. Parzen, *Time Series Analysis Papers*, Holden-Day, San Francisco, 1967, p. 542.
44. S. K. Srinivasan and S. Kumaraswamy, Stochastic Models for Fatigue Failure of Metals (to be published).
45. S. Kumaraswamy, Stochastic Processes and Continuum Mechanics, Ph.D. Thesis, Indian Institute of Technology, Madras, 1969.
46. S. R. Valluri, *Acta Metallurg.*, **11**(1963), 759.
47. R. W. Lardner, *J. Mech. Phys. Solids*, **14**(1966), 141, 281.

Chapter 6

STOCHASTIC PROBLEMS IN CONTINUUM MECHANICS

1. INTRODUCTION

In the last two chapters, we dealt with two major areas of applications of stochastic differential equations. Now we propose to discuss yet another area of application in the realm of Continuum Mechanics, where a variety of problems involving random structure occur in the basic differential equations describing the phenomena. Such problems have attracted the attention of a number of continuum theorists in different disciplines (see, for example, Beran [1], Lin [2]). We wish to confine our attention to a few of the many important problems that can be studied with the help of the techniques described in Chapters 2 and 3. In the choice of the material we have been mainly motivated by the importance from the viewpoint of stochastic theory in general rather than the technological role.

The organization of the chapter is as follows. Section 2 contains an account of random vibration of strings and rods. The vibration of nonlinear beams subject to random loading is then discussed. Section 4 deals with the behavior of columns. We then give a detailed account of stochastic flows in pipes and channels, the stochastic nature arising from either the boundary conditions or the applied pressure gradient. In Section 6 we describe a stochastic model for plug formation by impact.

2. RANDOM VIBRATIONS OF STRINGS AND RODS

In this section we consider two different problems. First we study the excitation of a nonlinear elastic string subject to random loading. In Section 2.2 we deal with the longitudinal vibrations of an elastic rod one of whose ends is subject to pulses randomly distributed in time.

2.1. Response of a Nonlinear String to Random Loading

Let us consider an elastic string (linear density ρ, elastic modulus E, length L) clamped at its ends and subjected to random loading. If the deflections and slopes are considered to be moderately small (consistent with the string retaining its elastic properties), the deflection $u(x, t)$ at any time t corresponding to a distance x measured from one of its ends is given by

$$\rho\frac{\partial^2 u}{\partial t^2} + \rho\beta\frac{\partial u}{\partial t} = \left[T_0 + \frac{AE}{L} \int_0^L \left(\frac{\partial u}{\partial x}\right)^2 dx \right]\frac{\partial^2 u}{\partial x^2} + f(x, t), \qquad (2.1)$$

where β is the viscous damping coefficient and A is the cross-sectional area of the string. The function $f(x, t)$ represents the random load. The boundary conditions are given by

$$u(0, t) = 0 = u(L, t). \qquad (2.2)$$

Caughey [3] studied this problem by assuming that $f(x, t)$ has a clipped Gaussian white spectrum described by

$$E\{f(x_1, t_1)f(x_2, t_2)\} = 4D\,\delta(x_1 - x_2)\frac{\sin \omega_c(t_1 - t_2)}{(t_1 - t_2)}. \qquad (2.3)$$

In the next few paragraphs we present the main conclusions of Caughey. Making use of the fact that the eigenmodes of (2.1) are essentially sine waves, we assume that $f(x, t)$ and $u(x, t)$ are given by

$$f(x, t) = \sum_{m=1}^{N} a_m \sin \frac{m\pi x}{L},$$

$$u(x, t) = \sum_{m=1}^{N} b_m \sin \frac{m\pi x}{L}, \qquad (2.4)$$

where N is chosen such that $\omega_N < \omega_c < \omega_{N+1}$. The justification for this will become apparent later.

We observe next that the power spectrum $g_i(\omega)$ of a_i is given by

$$g_i(\omega) = \frac{4D}{L\pi}, \qquad 0 < \omega < \omega_c,$$

$$= 0, \qquad \omega > \omega_c. \qquad (2.5)$$

Substituting (2.4) in (2.1), we find

$$\ddot{b}_n + \beta\dot{b}_n + \left(\frac{n\pi}{L}\right)^2 \frac{T_0}{\rho}\left(1 + \alpha \sum_{i=1}^{N} i^2 b_i^2\right)b_n = \frac{a_n}{\rho}, \qquad (2.6)$$

where

$$\alpha = \frac{AE}{4T_0 L}\left(\frac{\pi}{L}\right)^2. \qquad (2.7)$$

Caughey [3] has used the technique of quasilinearization where β is small by writing (2.6) in the form

$$\ddot{b}_n + \beta\dot{b}_n + \left(\frac{n\pi}{L}\right)^2 \frac{T_0}{\rho}K_n b_n + \varepsilon(b_1, b_2, \ldots, b_N) = \frac{a_n}{\rho}, \qquad (2.8)$$

where K_n is so chosen as to minimize $\varepsilon(b_1, b_2, \ldots, b_N)$. The minimization procedure consists in writing the time average of ε^2 and then estimating it as the stochastic average. With the neglect of ε, (2.8), in view of its linear nature, implies a Gaussian law for b_n. Using this result, Caughey has proved that the choice of K_n given by

$$K_n = 1 + \alpha \left[\sum_{i=1}^{N} i^2 \sigma_i^2 + 2n^2 \right] \sigma_n^2, \tag{2.9}$$

where $\sigma_n^2 = \mathbf{E}\{b_n^2\}$, minimizes ε. Using this value of K_n, $B_n(\omega)$, the power spectrum of b_n, is calculated:

$$B_n(\omega) = g_n(\omega)\rho^{-1} \left\{ \left[\left(\frac{n\pi}{L} \right)^2 \frac{T_0}{\rho} K_n - \omega^2 \right]^2 + \omega^2 \beta^2 \right\}^{-1}. \tag{2.10}$$

Thus the mean squared value of b_n is obtained by the well-known formula

$$\mathbf{E}\{b_n^2\} = \int_0^\infty B_n(\omega)\, d\omega. \tag{2.11}$$

We notice that, if β is small and $(n\pi/L)^2(T_0/\rho)K_n < \omega_c^2$,

$$\mathbf{E}\{b_n^2\} = \frac{2D}{\rho L} \left\{ \left(\frac{n\pi}{L} \right)^2 \frac{T_0}{\rho} K_n \beta \right\}^{-1}, \tag{2.12}$$

the main contribution to the integral coming from a small region close to

$$\omega^2 = \left(\frac{n\pi}{L} \right)^2 \frac{T_0}{\rho} K_n. \tag{2.13}$$

On the other hand, if $(n\pi/L)^2(T_0/\rho)K_n > \omega_c^2$, the integral (2.11) will be very small, thus justifying the approximate representation given by (2.4).

If K_n is substituted from (2.9), we have a system of simultaneous equations in σ_n^2 the solution of which can be found to be

$$\sigma_n^2 = \mathbf{E}\{b_n^2\} = \{ -(1 + zS) + [(1 + zS)^2 + 8z]^{\frac{1}{2}} \}(4n^2\alpha)^{-1}, \tag{2.14}$$

where

$$zS = \{ -1 + [1 + 4(N + z)z]^{\frac{1}{2}} \} \left[2\left(1 + \frac{2}{N} \right) \right]^{-1}, \tag{2.15}$$

$$z = \alpha \frac{2D}{\rho L} \left[\left(\frac{\pi}{L} \right)^2 \frac{T_0}{\rho} \beta \right]^{-1}. \tag{2.16}$$

Using equation (2.14), Caughey estimated the mean square deflections. It turns out that the mean square deflection in the nonlinear case is always less than the corresponding quantity in an equivalent linear string.

2.2. Aperiodic Vibrations of an Elastic Rod

In recent years much research effort has been devoted to the study of linear elastic structures subjected to random forces [4–6]. With a knowledge of the probability distribution of stress peaks in a randomly excited structure, it is possible to estimate the fatigue life of the structure. In this section we present some results (obtained by Subramanian and Kumaraswamy [7]) for an elastic beam subjected to a special type of forcing field.

Let us consider a rod of length l and cross-sectional area A fixed at one end (taken to be at $x = 0$) and acted on by a force $P(t)$ at the other end. It is well known that the aperiodic vibrations of the rod, assuming the homogeneous initial conditions

$$u(x, 0) = 0, \qquad \dot{u}(x, 0) = 0, \tag{2.17}$$

are determined by the equation

$$c^2\frac{\partial^2 u}{\partial x^2} - \frac{\partial^2 u}{\partial t^2} = 0 \tag{2.18}$$

with the boundary conditions

$$u(0, t) = 0, \qquad EA\frac{\partial u}{\partial x}\bigg|_l = P(t), \tag{2.19}$$

where $u(x, t)$ denotes the displacement and $c^2 = E/\rho$, E being the elasticity modulus, ρ the density. The solution of the (2.18) is given by

$$u(x, t) = \frac{2c}{\pi EA} \sum_{n=1}^{\infty} \frac{(-1)^{n-1}}{n - \frac{1}{2}} \sin(n - \tfrac{1}{2})\pi xl \cdot$$
$$\int_0^t P(\tau) \sin\frac{(2n-1)\pi c(t - \tau)}{2l}\, d\tau. \tag{2.20}$$

We assume $P(t)$ to be made up of a series of impulses of random strength $a(t_i)$ acting at random instants of time t_i. The displacement at any time t can be written as the stochastic integrals

$$u(x, t) = \frac{2c}{\pi EA} \sum_{n=1}^{\infty} \frac{(-1)^{n-1}}{n - \frac{1}{2}} \sin(n - \tfrac{1}{2})\pi xl \cdot$$
$$\int_0^t a(\tau) \sin\frac{(2n-1)\pi(t - \tau)}{2l}\, dN(\tau), \tag{2.21}$$

where $dN(\tau)$ is the random variable (represented as a point process) representing the number of pulses that occur in $(\tau, \tau + d\tau)$. The stress $\sigma(t)$ can be expressed as another stochastic integral by observing that

$$\sigma(x, t) = E\frac{\partial u(x, t)}{\partial x}. \tag{2.22}$$

By assuming that the processes $a(\tau)$ and $dN(\tau)$ are statistically independent, we can obtain the moments and correlations of $\sigma(x, t)$ and $u(x, t)$ as a weighted integral of product densities of the impulses defined on the time axis. Explicit expressions for the second-order correlation of displacement and stress are given in [7].

If we assume that the times of occurrence of the impulses are governed by the Poisson process, the probability frequency functions $\pi_1(u, t)$ and $\pi_2(\sigma, t)$ of $u(x, t)$ and $\sigma(x, t)$ can be obtained by the use of the Eq. (4.13) of Chapter 2. $\pi_1(u, t)$ and $\pi_2(\sigma, t)$ are given by

$$\pi_1(u, t) = \frac{1}{2\pi} \int_{-\infty}^{+\infty} \exp\left[- isu + \sum_{m=1}^{\infty} \frac{(is)^m}{m!} \eta_m(t) \right] ds,$$

$$\pi_2(\sigma, t) = \frac{1}{2\pi} \int_{-\infty}^{+\infty} \exp\left[- is\sigma + \sum_{m=1}^{\infty} \frac{(is)^m}{m!} \zeta_m(t) \right] ds, \qquad (2.23)$$

where

$$\eta_m(t) = \lambda \int_0^t h^m(\tau) \mathrm{E}\{a^m(\tau)\} \, d\tau,$$

$$\zeta_m(t) = \lambda \int_0^t g^m(\tau) \mathrm{E}\{a^m(\tau)\} \, d\tau,$$

$$h(t) = \frac{2c}{\pi E A} \sum_{n=1}^{\infty} \frac{(-1)^{n-1}}{n - \frac{1}{2}} \sin(n - \tfrac{1}{2}) \frac{\pi x}{l} \frac{\sin(2n - 1)\pi c t}{2l},$$

$$g(t) = \frac{2c}{IA} \sum_{n=1}^{\infty} (-1)^{n-1} \cos(n - \tfrac{1}{2}) \frac{\pi x}{l} \frac{\sin(2n - 1)\pi c t}{2l}. \qquad (2.24)$$

We shall assume that failure occurs when the fraction of time for which $\sigma > \sigma_F$ (σ_F is a certain fixed stress) is greater than some predetermined δ (see, for example, Crandall and Mark [8]). The mean time $\psi(\sigma_F)$ for which $\sigma(x, t)$ exceeds the threshold σ_F during the interval $(0, T)$ is given by

$$\psi(\sigma_F) = \int_0^T dt \int_{\sigma_F}^{\infty} d\sigma \pi_2(\sigma, t), \qquad (2.25)$$

so that failure can be expected to occur if $\psi(\sigma_F) > \delta$. The integral on the right-hand side of (2.25) can be evaluated numerically for given values of σ_F.

3. NONLINEAR BEAMS SUBJECT TO RANDOM EXCITATION

The response of linear continuous structures to random excitation was dealt with in Chapter 5. It is therefore quite natural to inquire whether nonlinear structures can be studied. As a simple example, we shall deal with elastic beams that are randomly excited. One of the important aspects of the problem is to study the effect of nonlinearity on the stresses. Although the displacements of a beam are considerably reduced, there is no reason to expect the same for the stresses (see, for example, Herbert [9]). Herbert [10] made a detailed study of a nonlinear Bernoulli–Euler beam subject to realistic random loading. In this section we present his results.

Let us consider an elastic beam with pin-ended supports which are restrained from motion. The equation governing moderately large vibrations is given by

$$EI\frac{\partial^4 w}{\partial x^4} + N\frac{\partial^2 w}{\partial x^2} + \rho A\frac{\partial^2 w}{\partial t^2} + \beta\frac{\partial w}{\partial t} = q(x, t), \qquad (3.1)$$

where the membrane force N is given by

$$N = \frac{EA}{2L} \int_0^L \left(\frac{\partial w}{\partial x}\right)^2 dx \qquad (3.2)$$

and E is the elastic modulus, I the moment of inertia of the cross section, ρ the mass density, A the cross-sectional area, L the length of the beam, and q the random load per unit length acting on the beam. Next we expand w and q in terms of the eigenfunctions of the linear problem:

$$w(x, t) = \sum w_m(t) \sin \frac{m\pi x}{L}, \qquad (3.3)$$

$$q(x, t) = \sum q_m(t) \sin \frac{m\pi x}{L}. \qquad (3.4)$$

Substituting (3.3) and (3.4) in (3.1), we obtain

$$\ddot{w}_m + \beta_0 \dot{w}_m + \omega_m^2 \left(1 + \frac{1}{4R^2 m^2} \sum_{n=1}^{\infty} n^2 w_n^2\right) w_m = a_m, \qquad (3.5)$$

where

$$\beta_0 = \frac{\beta}{\rho A}, \qquad a_m = \frac{q_m}{A}, \qquad \omega_m^2 = \frac{\pi^4 EI m^4}{AL^4}, \qquad R^2 = \frac{I}{A},$$

$$q_m(t) = \frac{2}{L} \int_0^L q(x, t) \sin \frac{m\pi x}{L} dx. \qquad (3.6)$$

Equation (3.5) represents a system of completed nonlinear stochastic differential equations with a random driving force $a_m(t)$. For small nonlinearities we can obtain approximations to the statistical properties of $w_m(t)$ by the method of equivalent linearization. Thus we can rewrite (3.5) in the form

$$\ddot{w}_m + \beta_0 \dot{w}_m + k_m \omega_m^2 w_m + \varepsilon_m = a_m, \qquad (3.7)$$

where

$$\varepsilon_m = \omega_m{}^2 w_m\left(1 - k_m + \frac{1}{4R^2 m^2}\sum_{n=1}^{\infty} n^2 w_n{}^2\right). \tag{3.8}$$

If we assume that the load is Gaussian, then (3.7), with the neglect of ε_m, implies a Gaussian distribution for w_n. The minimization of $E\{\varepsilon_m{}^2\}$ leads to the choice

$$k_m = 1 + \frac{1}{4R^2 m^2}\left(\sum_{n=1}^{\infty} n^2 \sigma_n{}^2 + 2m^2\sigma_m{}^2\right). \tag{3.9}$$

If we assume that

$$E\{q(x, t)q(y, t + \tau)\} = \delta(x - y)R(\tau), \tag{3.10}$$

where $R(\tau)$ satisfies certain feasibility conditions and if ε_m is neglected in (3.7), the mean square value of w_m is given by

$$\sigma_m{}^2 = E\{w_m{}^2\}$$

$$= \frac{2}{\rho^2 A^2 L}\int_{-\infty}^{+\infty}\int_{-\infty}^{+\infty} h_m(\tau_1)h_m(\tau_2)R(\tau_1 - \tau_2)\, d\tau_1\, d\tau_2 \tag{3.11}$$

with

$$h_m(\tau) = \frac{1}{2\pi}\int_{-\infty}^{+\infty} \frac{e^{i\omega\tau}\, d\omega}{\omega_m{}^2 k_m - \omega^2 + i\beta_0\omega}. \tag{3.12}$$

If the total power of the input is to be finite, it is clear that

$$\int_{-\infty}^{+\infty} R(\tau)\, d\tau < \infty.$$

A simple form having this feature is given by

$$R(\tau) = N_0 L\omega_0 e^{-|\omega_0\tau|}. \tag{3.13}$$

Equation (3.13) will yield the familiar bell-shaped curve centred around ω_0 in the frequency domain. On the basis of (3.13), $\sigma_m{}^2$ can be explicitly calculated:

$$\sigma_m{}^2 = \frac{N_0}{\rho^2 A^2}\frac{\omega_0 + (\omega_0{}^2/\beta_0)(1 + \omega_0{}^2/k_m\omega_m{}^2 - \beta_0{}^2/k_m\omega_m{}^2)}{(k_m\omega_m{}^2 + \omega_0{}^2)^2 - \beta_0{}^2\omega_0{}^2}. \tag{3.14}$$

If we specialize to the case of light damping, so that

$$\beta_0 \ll \omega_n, \qquad \beta_0 \ll \omega_0,$$

we obtain

$$\sigma_m{}^2 = \frac{\sigma_0{}^2}{k_m m^4(m^4 k_m + \mu)}, \tag{3.15}$$

where

$$\mu = \frac{\omega_0{}^2}{\omega_1{}^2}, \qquad \sigma_0{}^2 = \frac{N_0 L^4}{\beta E I \pi^4}. \tag{3.16}$$

Equations (3.15) and (3.9), taken along with the main equation (3.3) (with the neglect of ε_m), constitute the complete solution of the problem. However, the elimination of k_m is not such a simple process. An alternative approach is to compute k_m on the basis of the linear equation. In such a case, k_m is given by

$$k_m = 1 + \frac{\sigma_0^2}{4R^2}\left[\frac{F(\mu)}{m^2} + \frac{2\mu}{m^4(m^4 + \mu)}\right], \tag{3.17}$$

where

$$F(\mu) = \sum_{m=1}^{\infty} \frac{\mu}{m^2(m^4 + \mu)}. \tag{3.18}$$

On the basis of (3.17), Herbert [10] obtained numerical estimates of the total mean square value of the stress. His results indicate that the percentage reduction of the mean square stresses *can be substantially less than the percentage reduction of the mean square displacements.* Apart from this, Herbert also found that, as the spectral density of the load becomes wider, the difference in the percentage reduction of stress and displacement becomes greater.

4. BEHAVIOR OF COLUMNS

In the previous section we dealt with stochastic problems that arise in the study of nonlinear beams subject to random excitation. We now study the bending and buckling of columns. The problem is essentially linear and hence is capable of exact solution. Boyce [11, 12] made an extensive analysis of columns that are subject to random loadings as well as initial random displacements, and we present a brief account of his results.

4.1. Bending of Columns under Axial Loads

The classical theory of column buckling rests mainly on the following assumptions:

(i) The column is initially straight.

(ii) The loads are perfectly centered and parallel to the axis of the column.

(iii) The column is made of homogeneous isotropic elastic material.

(iv) The strains are small.

Under these conditions the lateral displacement $y(s)$ at distance s from an end satisfies the equation

$$\frac{d^4y}{ds^4} + \frac{P}{EI}\frac{d^2y}{ds^2} = 0, \qquad 0 < s < 1, \tag{4.1}$$

where EI is the flexural rigidity and P is the applied load. The boundary conditions are typically linear and homogeneous in y and its first three derivatives, thus leading to an eigenvalue problem for the critical loads $P_1, P_2, \ldots$ for which nontrivial displacements may occur. The smallest of these critical loads is commonly called the Euler (buckling) load, and no lateral displacements are predicted until P attains this value.

In engineering structures there is, to some extent, probably ignorance of the exact loads and material properties involved. Thus it is of interest to study the effects produced on the classical solution by random deviations from one or more of the assumptions describing the idealized mathematical load. Boyce [11] relaxed condition (ii) by assuming that the loads, though axial in direction, are eccentrically located with respect to the axis of the column. The boundary conditions are given by

$$y(0) = y(1) = 0, \tag{4.2}$$

$$\frac{d^2}{ds^2} y(0) = -\frac{P\varepsilon_1}{EI}, \qquad \frac{d^2}{ds^2} y(1) = -\frac{P\varepsilon_2}{EI}, \tag{4.2}$$

where ε_1 and ε_2 represent the eccentricities at $s = 0, 1$, respectively.

To analyze this problem we introduce the variables

$$x = \frac{s}{l}, \qquad Y = \frac{y}{h}, \qquad k^2 = \frac{Pl^2}{EI},$$

$$e_1 = \frac{\varepsilon_1}{h}, \qquad e_2 = \frac{\varepsilon_2}{h}. \tag{4.3}$$

into (4.1) and (4.2) so that we have

$$\frac{d^4Y}{dx^4} + \frac{k^2 d^2Y}{dx^2} = 0,$$

$$Y(0) = Y(1) = 0,$$

$$\frac{d^2}{dx^2} Y(0) = -k^2 e_1, \qquad \frac{d^2}{dx^2} Y(1) = -k^2 e_2. \tag{4.4}$$

The solution of this equation can be written in the form

$$Y(x) = e_1 f(k, 1 - x) + e_2 f(k, x), \tag{4.5}$$

where

$$f(k, x) = \frac{\sin kx}{\sin k} - x.$$

The preceding equation for Y and its solution are of the type discussed in Section 5 of Chapter 3. The p.f.f. of Y can be obtained from the joint p.f.f.

of e_1 and e_2 by suitable transformation of variables. Boyce has explicitly obtained all the essential statistical characteristics of $Y(x)$.

4.2. Buckling of Columns with Random Initial Displacements

In Section 4.1 we studied the implications of the deviation of the classical model of a column arising from the load being not perfectly centered. We assumed that the column is initially straight. In engineering design it is not always possible to meet this requirement. It is eminently reasonable to assume small random variations of the centerline and, in such a case, (4.4) takes the form

$$\frac{d^4Y}{dx^4} + k^2\frac{d^2Y}{dx^2} = -k^2\frac{d^2w}{dx^2}, \tag{4.6}$$

where $w(x)$ is the initial random displacement of the centerline. Well-behaved nontrivial solutions exist for all values of k except the eigenvalues. The analysis of this problem can be carried out under any given reasonable boundary conditions. If, for example, we assume that both ends are simply supported so that

$$Y(0) = Y(1) = Y''(0) = Y''(1) = 0, \tag{4.7}$$

then the differential equation can be integrated twice, leading to the equation

$$\frac{d^2Y}{dx^2} + k^2Y = -k^2w, \qquad 0 < x < 1, \tag{4.8}$$

an equation of the type discussed in Section 2 of Chapter 3. If $w(x)$ represents a nonstationary shot noise, then the characteristic functional can be obtained from the principal result, (4.9) of Chapter 2. If, on the other hand, $w(x)$ represents a Gaussian (but not necessarily of white noise type), then $w(x)$ can be expressed as an appropriate weighted integral of a white noise process so that the statistical properties can be deduced directly by the methods explained in Section 2 of Chapter 3. Boyce studied the statistical characteristics of the solution by writing Y explicitly as a linear functional of w. He obtained the first two moments of Y for a general $w(x)$ and also the probability that the absolute midspan displacement lies between given bounds.

5. STOCHASTIC FLOWS THROUGH PIPES AND CHANNELS

The motion of a viscous, homogeneous incompressible fluid in circular pipes of uniform cross section has been extensively studied by many workers. The flow due to a constant pressure gradient in the stationary case is the well-known Poiseuille flow for which the velocity along the axis of the pipe has parabolic distribution with respect to the ordinate perpendicular to the axis of the pipe. Some years ago Lance [13] made an analysis of the flow of a viscous incompressible fluid in a circular pipe and in a two-dimensional channel subjected to a series of pulses (deterministic in nature) parallel to the flow direction. He concluded that, when the pulses act in the opposite direction to the pressure gradient, the total flow can be stopped under certain conditions. This phenomenon would perhaps explain the failure, due to fuel starvation, of jet engines fired at. A complete study of the problem can be made by assuming that the strength of the impulses and their times of occurrences are completely random in nature. Equivalently, as will be clear presently, we can also assume a random pressure gradient without any random conditions on the boundary. We present in this section a systematic analysis of the problem.

Let us consider a fluid of constant density ρ and viscosity μ contained in a long horizontal pipe of length L and radius R, the fluid initially being in a state of rest. We assume that the pipe is subjected to a series of pulses in the axial direction which do not appreciably move the pipe. The characteristics of the flow in a pipe (as well as in a channel) have been determined by Srinivasan, Subramanian, and Kumaraswamy [14] when the pressure gradient is a random function of time and the series of pulses are also of random strength, acting at random instants of time. In the next few subsections, we summarize the main results in this direction.

5.1. Flow in a Pipe of Circular Cross Section

At time $t = 0$ a pressure gradient (which is a random function) is set up, the system being initially at rest. In addition, the pipe is subjected to pulses of random strength acting at random instants of time. We assume that the motion is only along the axis of the pipe which we take as the z-axis. The equations of motion in cylindrical coordinates reduce to the fact that p is a function of z and t only, and

$$\frac{\partial w}{\partial t} - \frac{\mu}{\rho}\left[\frac{\partial^2 w}{\partial r^2} + \frac{1}{r}\frac{\partial w}{\partial r}\right] = -\frac{1}{\rho}\frac{\partial p}{\partial z}, \tag{5.1}$$

where w, the velocity along the z-axis, is a function of r and t, a consequence of circular symmetry and the equation of continuity. Equation (5.1) takes the simplified form

$$\frac{\partial^2 v}{\partial \xi^2} + \frac{1}{\xi}\frac{\partial v}{\partial \xi} = -F(\tau) + \frac{\partial v}{\partial \tau} \tag{5.2}$$

by introducing the dimensionless variables ξ, τ, and v given by

$$\xi = \frac{r}{R}, \qquad \tau = \frac{\mu t}{\rho R^2},$$

$$w = u_0 v, \qquad \frac{\partial p}{\partial z} = -\frac{\mu u_0}{R^2}F(\tau). \tag{5.3}$$

Thus (5.2) is a stochastic partial differential equation subject to the following conditions:

$$v = 0, \quad \text{at } \tau = 0,$$
$$v = \sum A_i \, \delta(\tau - \tau_i) = f(\tau), \quad \text{at } \xi = 1, \tag{5.4}$$
$$v(0, \tau) = 0, \quad \tau < 0, \tag{5.5}$$
$$|v| < \infty \quad \text{and} \quad \left|\frac{\partial v}{\partial \xi}\right| < \infty \quad \text{at } \xi = 0, \quad \xi = 1. \tag{5.6}$$

A_i in (5.5) is a nondimensional random quantity giving the strength of the ith pulse, and τ_i is a random instant at which the ith pulse occurs.

Multiplying (5.2) by $\xi J_0(\alpha_i \xi)$ (α_i being a root of $J_0(x_i) = 0$) and integrating over ξ from 0 to 1, we get

$$\frac{\partial}{\partial \tau} \bar{v}(\alpha_i, \tau) + \alpha_i^2 \bar{v}(\alpha_i, \tau) = \frac{1}{\alpha_i} F(\tau) J_1(\alpha_i) + f(\tau) \alpha_i J_1(\alpha_i), \tag{5.7}$$

where $\bar{v}$ is the finite Hankel transform of v defined by

$$\bar{v}(\alpha_i, \tau) = \int_0^1 \xi J_0(\alpha_i \xi) v(\xi, \tau) \, d\xi. \tag{5.8}$$

Solving (5.7), we obtain

$$\bar{v}(\alpha_i, \tau) = \int_0^\tau \phi(\tau') \exp[-\alpha_i^2(\tau - \tau')] \, d\tau' \tag{5.9}$$

where

$$\phi(\tau') = \alpha_i J_1(\alpha_i) f(\tau') + \frac{F(\tau')}{\alpha_i} J_1(\alpha_i). \tag{5.10}$$

Using the inversion theorem, we obtain

$$v(\xi, \tau) = 2 \sum_i \frac{J_0(\alpha_i \xi)}{J_1^2(\alpha_i)} \bar{v}(\alpha_i, \tau), \tag{5.11}$$

where the summation extends over all the positive roots of the equation $J_0(x) = 0$,

$$v(\xi, \tau) = 2 \sum_i \frac{\alpha_i J_0(\alpha_i \xi)}{J_1(\alpha_i)} \left[\sum_k A_k e^{-\alpha_i^2(\tau - \tau_k)} H(\tau - \tau_k) \right] +$$
$$2 \sum_i \frac{J_0(\alpha_i \xi)}{J_1(\alpha_i)} \frac{1}{\alpha_i} \int_0^\tau F(\tau') e^{-\alpha_i^2(\tau - \tau')} \, d\tau', \tag{5.12}$$

where H is the Heaviside function.

The quantity of fluid flowing (relative to the pipe) through a unit length of the pipe is given by

$$Q(\tau) = \int_0^1 v(\xi, \tau) 2\pi\xi \, d\xi$$

$$= 4\pi \sum_i \left\{ \sum_k A_k e^{-\alpha_i^2(\tau - \tau_k)} H(\tau - \tau_k) + \frac{1}{\alpha_i^2} \int_0^\tau F(\tau') e^{-\alpha_i^2(\tau - \tau')} \, d\tau' \right\}. \quad (5.13)$$

The velocity $v(\xi, \tau)$ given by (5.12) can be written in terms of the stochastic integral:

$$v(\xi, \tau) = 2\sum_i \frac{\alpha_i J_0(\alpha_i \xi)}{J_1(\alpha_i)} \int_0^\tau A(t) e^{-\alpha_i^2(\tau - t)} \, dN(t) +$$

$$2\sum_i \frac{J_0(\alpha_i \xi)}{\alpha_i J_1(\alpha_i)} \int_0^\tau F(t) e^{-\alpha_i^2(\tau - t)} \, dt, \quad (5.14)$$

where $N(t)$ represents the number of pulses in time t, so that $dN(t)$ represents the number of pulses in the interval $(t, t + dt)$. The moments of the distribution of $v(\xi, \tau)$ can be obtained with the help of the product densities (see Ramakrishnan [15]) introduced in Chapter 5. For the purpose of our analysis in this section, we assume that $F(t)$ is a deterministic function of t, say of the form ae^{-bt}. The points of the time axis τ_i corresponding to the times of occurrence of pulses are assumed to be distributed in accordance with the Poisson law with constant average density λ. The product densities in this case are

$$\phi_1(t) = \lambda, \qquad \phi_2(t_1, t_2) = \lambda^2, \quad (5.15)$$

We then impose the following restrictions on the distribution of $A(t_i)$:

$$\mathrm{E}\{A(t)\} = I, \qquad \text{a constant,}$$
$$\mathrm{E}\{A(t_1)A(t_2)\} = \psi(t_1)e^{-\mu(t_2 - t_1)}, \qquad \mu > 0, \qquad t_2 > t_1. \quad (5.16)$$

These assumptions imply that the random process considered is nonstationary in character. The expressions $\mathrm{E}\{v(\xi, \tau)\}$ and $\mathrm{E}\{Q(\tau)\}$ are given by

$$\mathrm{E}\{v(\xi, \tau)\} = 2\sum_i \frac{I J_0(\alpha_i \xi)}{J_1(\alpha_i)} \frac{1 - e^{-\alpha_i^2}}{\alpha_i} + 2\sum_i \frac{a J_0(\alpha_i \xi)}{\alpha_i J_1(\alpha_i)} \frac{e^{-b\tau} - e^{-\alpha_i^2 \tau}}{\alpha_i^2 - b} \quad (5.17)$$

and

$$\mathrm{E}\{Q(\tau)\} = 4\pi \sum_i \left[\frac{I}{\alpha_i^2}(1 - e^{-\alpha_i^2 \tau}) + \frac{a}{\alpha_i^2} \frac{e^{-b\tau} - e^{-\alpha_i^2 \tau}}{\alpha_i^2 - b} \right]. \quad (5.18)$$

In the particular case when the process governing the pulses is stationary, ψ in (5.16) will be a constant. Explicit expressions for the second-order correlation of velocity as well as discharge have been derived by Srinivasan, Subramanian, and Kumaraswamy [14].

The flow is reversed in an average sense if the expected value of $Q(\tau)$ is negative, the condition for this being

$$\sum_i \int_0^\tau E\{A(t)\}e^{-\alpha_i^2(\tau-t)}\, E\{dN(t)\} + \frac{1}{\alpha_i^2} \int_0^\tau E\{F(t)\}e^{-\alpha_i^2(\tau-t)}\, dt < 0. \qquad (5.19)$$

In the case when the distributions of $A(t)$, $dN(t)$, are as prescribed in (5.15) and (5.16) and $F(t)$ is deterministic and equal to ae^{-bt}, the condition above reduces to

$$\sum_i \left[I\lambda(1 - e^{-\alpha_i^2\tau}) + \frac{a}{\alpha_i^2 - b}(e^{-b\tau} - e^{-\alpha_i^2\tau}) \right] < 0. \qquad (5.20)$$

Since the pulses act in a direction opposing the free stream, I is negative.

5.2. Flow in a Pipe Subjected to a Random Pressure Gradient

We will assume the boundary condition at $\xi = 1$ to be

$$v(1, \tau) = 0, \qquad \tau > 0; \qquad (5.21)$$

that is, the pipe is not subjected to pulses. We shall, however, assume that the $F(\tau)$ characterizing the pressure gradient is a random function of τ and study the resulting velocity correlation. The velocity is then given by

$$v(\xi, \tau) = 2\sum_i \frac{J_0(\alpha_i\xi)}{\alpha_i J_1(\alpha_i)} \int_0^\tau F(t)e^{-\alpha_i^2(\tau-t)}\, dt. \qquad (5.22)$$

The expected value and the velocity correlation at two different points distant ξ_1, ξ_2 from the axis of the pipe at two different instants of time are given by

$$E\{v(\xi, \tau)\} = 2\sum_i \frac{J_0(\alpha_i\xi)}{\alpha_i J_1(\alpha_i)} \int_0^\tau E\{F(t)\}e^{-\alpha_i^2(\tau-t)}\, dt, \qquad (5.23)$$

$$E\{v(\xi_1, \tau_1)v(\xi_2, \tau_2)\} =$$
$$4\sum_{i,j} \int_0^{\tau_1} \int_0^{\tau_2} E\{F(t_1)F(t_2)\}\frac{J_0(\alpha_i\xi_1)J_0(\alpha_j\xi_2)}{J_1(\alpha_i)J_1(\alpha_j)\alpha_i\alpha_j} \cdot$$
$$e^{-\alpha_i^2(\tau_1-t_1)}\, e^{-\alpha_j^2(\tau_2-t_2)}\, dt_1\, dt_2. \qquad (5.24)$$

From (5.24) we can get the correlation at the same point at different times and the correlation at different points at the same time. On the other hand, the mean square value of the velocity is given by

$$E\{v^2(\xi, \tau)\} =$$
$$4\sum_{i,j} \frac{J_0(\alpha_i\xi)J_0(\alpha_j\xi)}{J_1(\alpha_i)J_1(\alpha_j)\alpha_i\alpha_j} \int_0^\tau \int_0^\tau E\{F(t_1)F(t_2)\}e^{-\alpha_i^2(\tau-t_1)-\alpha_j^2(\tau-t_2)}\, dt_1\, dt_2. \qquad (5.25)$$

Assuming that

$$E\{F(t)\} = k, \qquad \text{a constant,} \tag{5.26}$$

$$E\{F(t_1)F(t_2)\} = (\sin \beta t_1)\, \delta(t_2 - t_1), \tag{5.27}$$

we notice that the variance $\sigma^2(v)$ is given by

$$\sigma^2(v) = E\{v^2\} - [E\{v\}]^2$$

$$= 4\sum_{i,j} \frac{J_0(\alpha_i \xi)J_0(\alpha_j \xi)}{J_1(\alpha_i)J_1(\alpha_j)\alpha_i\alpha_j} \frac{e^{-(\alpha_i^2+\alpha_j^2)\tau}}{(\alpha_i^2 + \alpha_j^2) + \beta^2} \cdot$$

$$[e^{(\alpha_i^2+\alpha_j^2)\tau} \{(\alpha_i^2 + \alpha_j^2)(\sin \beta\tau) - \beta(\cos \beta\tau)\} + \beta] -$$

$$4k^2 \sum_{i,j} \frac{J_0(\alpha_i \xi)J_0(\alpha_j \xi)}{\alpha_i^3\alpha_j^3 J_1(\alpha_i)J_1(\alpha_j)}(1 - e^{-\alpha_i^2\tau})(1 - e^{-\alpha_j^2\tau}). \tag{5.28}$$

If the process $F(t)$ is stationary so that the second-order correlation of $F(t)$ depends only on the difference $\tau_2 - \tau_1 = \tau > 0$, then we have

$$\lim_{\substack{\tau_1 \to \infty,\, \tau_2 \to \infty, \\ \tau_2 - \tau_1 = \tau}} E\{v(\xi, \tau_1)v(\xi, \tau_2)\} =$$

$$4\sum_{i,j} \frac{J_0(\alpha_i \xi)J_0(\alpha_j \xi)}{J_1(\alpha_i)J_1(\alpha_j)\alpha_i\alpha_j}\left[\frac{1}{\alpha_i^2 + \alpha_j^2}\int_0^\infty R(t)e^{-\alpha_i^2|\tau - t| - \alpha_j^2(\tau + t)}\, dt\right]. \tag{5.29}$$

When $R(t) = e^{-\beta t}$, the power spectrum $P(\omega)$ is given by

$$P(\omega) = \frac{4}{\pi}\sum_{i,j} \frac{J_0(\alpha_i \xi)J_0(\alpha_j \xi)}{J_1(\alpha_i)J_1(\alpha_j)} \frac{2\alpha_i\beta}{\alpha_j(\alpha_i^2 + \alpha_j^2)(\alpha_i^4 + \omega^2)(\beta^2 + \omega^2)}. \tag{5.30}$$

5.3. Flow in a Straight-Walled Two-Dimensional Channel

Let x be the coordinate measured along the channel length, $2a$ the distance between the walls, and the origin midway between them. We assume that the components of the velocity in the y- and z-directions are zero:

$$u_y = u_z = 0. \tag{5.31}$$

The equation of continuity demands that the velocity component u_x, along the axis of the channel, should be a function of y and t only so that

$$u_x = u(y, t). \tag{5.32}$$

The equations of motion are

$$\rho\frac{\partial u}{\partial t} = -\frac{\partial p}{\partial x} + \mu\frac{\partial^2 u}{\partial y^2},$$

$$0 = -\frac{\partial p}{\partial y},$$

$$0 = -\frac{\partial p}{\partial z}. \tag{5.33}$$

The last two equations imply that p is a function of x and t only. The first equation becomes

$$\frac{\partial u}{\partial t} - \nu \frac{\partial^2 u}{\partial y^2} = -\frac{1}{\rho} \frac{\partial p}{\partial x}. \tag{5.34}$$

We note that (5.34) will hold if each member is equal to a function of t only. Introducing the dimensionless variables ξ, τ, and v given by

$$y = a\xi, \qquad \tau = \frac{\mu t}{\rho a^2}, \qquad u = u_0 v,$$

$$\frac{\partial p}{\partial x} = \frac{\mu u_0}{a^2} F(\tau), \tag{5.35}$$

note that (5.34) can be written

$$\frac{\partial^2 v}{\partial \xi^2} = \frac{\partial v}{\partial \tau} + F(\tau) \tag{5.36}$$

with the conditions

$$v, \frac{\partial v}{\partial \xi} \qquad \text{finite at } \xi = \pm 1,$$

$$v(0, \tau) = 0 \quad \text{for} \quad \tau < 0,$$

$$v(\pm 1, \tau) = \sum_{i=0}^{N} A_i \,\delta(\tau - \tau_i) \quad \tau > 0,$$

$$v = 0 \quad \text{at } \tau = 0, \tag{5.37}$$

where A_i is a nondimensional random quantity giving the strength of the ith pulse and τ_i is the random instant at which the ith pulse occurs. We put

$$2\eta = \xi + 1 \tag{5.38}*$$

and apply the Fourier finite sine transform technique:

$$\frac{\partial \bar{v}(n, \tau)}{\partial \tau} + \frac{\pi^2 n^2}{4} \bar{v}(n, \tau) = \left[\frac{1 - (-)^n}{n\pi} \right] \left[\frac{n^2 \pi^2}{4} f(\tau) - F(\tau) \right], \tag{5.39}$$

where

$$\bar{v}(n, \tau) = \int_0^1 v(\eta, \tau) \sin n\pi\eta \, d\eta$$

and n is a positive integer.

Thus the solution of (5.39) subject to the conditions (5.37) is given by

$$\bar{v}(n, \tau) = (1 - \cos n\pi)\frac{n\pi}{4} \left[\sum_i A_i e^{-(\pi^2 n^2/4)(\tau - \tau_i)} H(\tau - \tau_i) \right] -$$

$$\frac{1 - \cos n\pi}{n\pi} \int_0^\tau F(\tau') e^{-(\pi^2 n^2/4)(\tau - \tau')} \, d\tau'. \tag{5.40}$$

* We use the same functional symbol v to denote the velocity as a function of η also.

On using the inversion theorem we obtain

$$v(\eta, \tau) = 2 \sum_{m=0}^{\infty} \left\{ \sum_i \frac{A_i}{2}(2m + 1)\pi e^{-\pi^2(m+\frac{1}{2})^2(\tau-\tau_i)} H(\tau - \tau_i) - \right.$$
$$\left. \frac{2}{(2m + 1)\pi} \int_0^{\tau} F(\tau')e^{-\pi^2(m+\frac{1}{2})^2(\tau-\tau')} d\tau' \right\} \sin(2m + 1)\pi\eta. \quad (5.41)$$

The quantity of fluid flowing (relative to the channel) per unit width of the channel is given by

$$Q(\tau) = \int_{-1}^{+1} v(\xi, \tau) \, d\xi$$
$$= 2 \int_0^1 v(\eta, \tau) \, d\eta$$
$$= \sum_{m=0}^{\infty} \left\{ 4\sum_i A_i e^{-\pi^2(m+\frac{1}{2})^2(\tau-\tau_i)} H(\tau - \tau_i) - \right.$$
$$\left. \frac{16}{(2m + 1)^2\pi^2} \int_0^{\tau} F(\tau')e^{-\pi^2(m+\frac{1}{2})^2(\tau-\tau')} d\tau' \right\}. \quad (5.42)$$

The discharge can be written as before as the stochastic integral:

$$Q(\tau) = \sum_{m=0}^{\infty} \left[4 \int_0^{\tau} A(t)e^{-\pi^2(m+\frac{1}{2})^2(\tau-t)} \, dN(t) - \right.$$
$$\left. \frac{16}{(2m + 1)^2\pi^2} \int_0^{\tau} F(t)e^{-\pi^2(m+\frac{1}{2})^2(\tau-t)} \, dt \right], \quad (5.43)$$

where $N(t)$ represents the number of pulses in time t, so that $dN(t)$ represents the number of pulses in the interval $(t, t + dt)$. The moments of the distribution of $Q(\tau)$ can be studied with the help of product densities as before. The flow is reversed in an average sense if the expected value of $Q(\tau)$ is negative, the condition for this being

$$\sum_{m=0}^{\infty} 4 \int_0^{\tau} E\{A(t)\}e^{-\pi^2(m+\frac{1}{2})^2(\tau-t)} E\{dN(t)\} <$$
$$\sum_{m=0}^{\infty} \frac{16}{(2m + 1)^2\pi^2} \int_0^{\tau} E\{F(t)\}e^{-\pi^2(m+\frac{1}{2})^2(\tau-t)} \, dt. \quad (5.44)$$

In the case when the distributions of $A(t)$, $dN(t)$ are prescribed in (5.15) and (5.16), and $F(t)$ is deterministic and equal to ae^{-bt}, the condition can be written

$$\sum 4I\lambda \left[\frac{1 - e^{-\pi^2(m+\frac{1}{2})^2\tau}}{\pi^2(m + \frac{1}{2})^2} \right] < \sum \frac{16}{(2m + 1)^2\pi^2} \frac{e^{-b\tau} - e^{-\pi^2(m+\frac{1}{2})^2\tau}}{(2m + 1)^2\pi^2 - b}, \quad (5.45)$$

where I is negative as before.

5.4. Flow in a Channel Subject to a Random Pressure Gradient

We assume the boundary condition at $\xi = \pm 1$ to be

$$v(\pm 1, \tau) = 0, \qquad \tau > 0, \tag{5.46}$$

so that the walls of the channel are not subjected to any pulses. However, $F(\tau)$, which characterizes the pressure gradient, is a random function of τ. Proceeding as in Section 5.3, the mean value of the velocity $v(\eta, \tau)$ is given by

$$E\{v(\eta, \tau)\} = -\sum_m \frac{2}{M^{\frac{3}{2}}} \sin(2m + 1)\pi\eta \int_0^\tau E\{F(t)\}e^{-M(\tau-t)} \, dt, \tag{5.47}$$

where $M = \pi^2(m + \frac{1}{2})^2$.

If $E\{F(t)\} = K$, a constant, then, allowing $\tau \to \infty$, we get the steady state solution as

$$\lim_{\tau \to \infty} E\{v(\eta, \tau)\} = \frac{K}{2}(1 - \xi^2). \tag{5.48}$$

The correlations at two different points distant η_1 and η_2 from the $\eta = 0$ plane at two different instants of time can be calculated.

If $F(t)$ is a stationary process with a stationary correlation $R(t)$, the power spectrum $P(\omega)$ can be computed:

$$P(\omega) = \frac{16}{\pi} \sum_{m=0}^\infty \sum_{m=0}^\infty \frac{\sin(2m + 1)\,\pi\eta \sin(2n + 1)\,\pi\eta}{(2m + 1)(2n + 1)}$$

$$\left[\frac{(m + \frac{1}{2})^2}{\omega^2 + (m + \frac{1}{2})^2\pi^4} \int_0^\infty R(t)\,(\cos \omega t + e^{-\pi^2(m+\frac{1}{2})^2 t}) \, dt + \right.$$

$$\left. \frac{(n + \frac{1}{2})^2}{\pi(n + \frac{1}{2})^2 + \omega^2} \int_0^\infty R(t)e^{-\pi^2(n+\frac{1}{2})^2 t} \, dt \right]. \tag{5.49}$$

For the particular case when $R(t) = e^{-\beta t}$, (5.49) has been numerically evaluated for various values of η in [14]. The series is fastly convergent, and it is sufficient to include only the first nine terms to attain an accuracy up to four decimal places.

6. A STOCHASTIC MODEL FOR AN IMPACT PROBLEM

During the last few years increasing interest has been evinced in the solution of problems concerned with the impact of solid bodies. In particular, the subject of penetration and perforation of solid targets by projectiles is of considerable interest in ballistic research. Quite recently, Pytel and Davids [16] suggested a viscous model and studied the behavior of a plate subjected to projectile impact under conditions leading to plug-type failure. The impact was represented by a velocity uniformly distributed over a circular area on the plate surface, the problem being assumed to be axisymmetric. But the following considerations lead us to assume the initial velocity distribution to be a random function:

(i) The thickness of the plate may vary from point to point in a random manner.

(ii) The shapes of the bullets may not be identical.

(iii) The Young's modulus and the Poisson's ratio may vary randomly because of the inhomogeneity in the material, resulting in the random variations of the stiffness of the plate.

Motivated by these considerations, Subramanian and Kumaraswamy [17] analyzed this problem by studying the response of a plate subjected to projectile impacts, the impact being represented by a general random velocity distribution. Although the natural way of formulating this problem would be a three-dimensional one, these authors assumed axisymmetry to avoid computational difficulties. For the material response, a Voight-type behavior is assumed and all stresses are neglected, save the vertical shearing stress, which is assumed to depend only on the radial coordinates. In the next few subsections we present results obtained by these authors.

6.1. Formulation and Solution of the Problem

Let us consider an infinite plate of thickness h defined in a cylindrical coordinates system $0 < z < h$. The material of the plate is assumed to exhibit a Voight-type response. It is also supposed that the plate is set in motion by an initial random velocity $f(r)$ due to the impact and distributed over the surface area of the plate. We assume that all the quantities involved are independent of z; then the initial conditions can be written

$$w(r, 0) = 0, \qquad v(r, 0) = f(r), \tag{6.1}$$

w being the displacement in the z-direction and v the velocity in the same direction (equal to $\partial w/\partial t$).

The problem is axially symmetric and the equation of motion is

$$\frac{\partial \sigma}{\partial r} + \frac{1}{r}\sigma = \rho\frac{\partial v}{\partial t}, \tag{6.2}$$

where ρ is the density of the material of the plate and σ is the vertical shearing stress. Since the material is assumed to behave like a Voight substance, the stress is given by

$$\sigma = a\frac{\partial v}{\partial r} + b\frac{\partial w}{\partial r}. \tag{6.3}$$

In view of (6.3), Eq. (6.2) takes the form

$$a\left[\frac{\partial^2 v}{\partial r^2} + \frac{1}{r}\frac{\partial v}{\partial r}\right] + b\left[\frac{\partial^2 w}{\partial r^2} + \frac{1}{r}\frac{\partial w}{\partial r}\right] = \rho\frac{\partial v}{\partial t}. \qquad (6.4)$$

Taking the infinite Hankel transform with respect to r on both sides of (6.4), we have

$$-\xi^2[a\bar{v}(\xi, t) + b\bar{w}(\xi, t)] = \rho\frac{\partial\bar{v}}{\partial t}, \qquad (6.5)$$

where $\bar{v}$ and $\bar{w}$ are the infinite Hankel transforms of v and w, respectively, given by

$$\bar{v} = \int_0^\infty rJ_0(\xi r)v(r, t)\,dr,$$

$$\bar{w} = \int_0^\infty rJ_0(\xi r)w(r, t)\,dr. \qquad (6.6)$$

Since $\bar{v} = d\bar{w}/dt$, (6.5) can be rewritten:

$$\rho\frac{d^2\bar{w}}{dt^2} + a\xi^2\frac{d\bar{w}}{dt} + b\xi^2\bar{w} = 0, \qquad (6.7)$$

the solution of which is given by

$$\bar{w} = A(\xi)e^{m_1(\xi)t} + B(\xi)e^{m_2(\xi)t}, \qquad (6.8)$$

where

$$m_{1,2}(\xi) = \frac{-a\xi^2 \pm [a^2\xi^4 - 4b\rho\xi^2]^{\frac{1}{2}}}{2\rho} \qquad (6.9)$$

and $A(\xi)$, $B(\xi)$ are arbitrary functions of ξ. Therefore

$$\bar{v} = \frac{d\bar{w}}{dt} = m_1Ae^{m_1t} + m_2Be^{m_2t}.$$

Using the initial conditions, the arbitrary functions A and B are determined as

$$A = \frac{\bar{f}}{-(m_2 - m_1)} = -B. \qquad (6.10)$$

Making use of the inversion theorem for the Hankel transform, we obtain

$$w(r, t) = \int_0^\infty \bar{w}\xi J_0(\xi r)\,d\xi$$

$$= \int_0^\infty \xi J_0(\xi r)\frac{e^{m_1t} - e^{m_2t}}{m_1 - m_2}\bar{f}\,d\xi, \qquad (6.11)$$

$$v(r, t) = \int_0^\infty \xi J_0(\xi r)\frac{m_1e^{m_1t} - m_2e^{m_2t}}{m_1 - m_2}\bar{f}\,d\xi. \qquad (6.12)$$

The shearing stress σ is given by

$$\sigma = a\frac{\partial v}{\partial r} + b\frac{\partial w}{\partial r} = - \int_0^\infty \frac{\xi^2 J_1(\xi r)\bar{f}}{m_1 - m_2}[(am_1 + b)e^{m_1 t} - (am_2 + b)e^{m_2 t}]\, d\xi.$$

(6.13)

The expected values of the displacement, the velocity, and the shearing stress are given by

$$\mathbf{E}\{w(r, t)\} = \int_0^\infty \xi J_0(\xi r)\frac{e^{m_1 t} - e^{m_2 t}}{m_1 - m_2}\mathbf{E}\{\bar{f}\}\, d\xi,$$

(6.14)

$$\mathbf{E}\{v(r, t)\} = \int_0^\infty \xi J_0(\xi r)\frac{m_1 e^{m_1 t} - m_2 e^{m_2 t}}{m_1 - m_2}\mathbf{E}\{\bar{f}\}\, d\xi,$$

(6.15)

$$\mathbf{E}\{\sigma(r, t)\} = - \int_0^\infty \xi^2 J_1(\xi r)\frac{[e^{m_1 t}(am_1 + b) - e^{m_2 t}(am_2 + b)]}{m_1 - m_2}\mathbf{E}\{\bar{f}\}\, d\xi.$$

(6.16)

The second-order correlation of the displacement $w(r, t)$ is given by

$$\mathbf{E}\{w(r_1, t)w(r_2, t)\} = \int_0^\infty \int_0^\infty \frac{\xi\eta J_0(r_1\xi)J_0(r_2\xi)\mathbf{E}\{\bar{f}(\xi)\bar{f}(\eta)\}}{[m_1(\xi) - m_2(\xi)][m_1(\eta) - m_2(\eta)]} \cdot$$

$$[e^{[m_1(\xi)+m_1(\eta)]t} + e^{[m_2(\xi)+m_2(\eta)]t} -$$

$$e^{[m_1(\xi)+m_2(\eta)]t} - e^{[m_1(\eta)+m_2(\xi)]t}]\, d\xi\, d\eta.$$

(6.17)

The mean square value of $w(r, t)$ is obtained from (6.17) by putting $r_1 = r_2 = r$. Similarly, the corresponding expressions for $v(r, t)$ can easily be written.

Let us confine our attention to the case when the material response is of a purely viscous nature. This is a particular case of the Voight model where $a = v$ and $b = 0$. Thus we have the relations

$$m_1 = 0, \qquad m_2 = - \frac{v\xi^2}{\rho},$$

(6.18)

$$\mathbf{E}\{w(r, t)\} = \frac{\rho}{v} \int_0^\infty \frac{J_0(\xi r)}{\xi}[1 - e^{-v\xi^2 t/\rho}]\mathbf{E}\{\bar{f}(\xi)\}\, d\xi,$$

(6.14a)

$$\mathbf{E}\{v(r, t)\} = \int_0^\infty \xi J_0(\xi r)e^{-v\xi^2 t/\rho}\, \mathbf{E}\{\bar{f}(\xi)\}\, d\xi,$$

(6.15a)

$$\mathbf{E}\{\sigma(r, t)\} = - v \int_0^\infty \xi^2 J_1(r\xi)e^{-v\xi^2 t/\rho}\, \mathbf{E}\{\bar{f}(\xi)\}\, d\xi,$$

(6.16a)

$$\mathbf{E}\{w(r_1, t)w(r_2, t)\} = \frac{\rho^2}{v^2} \int_0^\infty \int_0^\infty \frac{J_0(\xi r_1)J_0(\eta r_2)}{\xi} \cdot$$

$$\mathbf{E}\{\bar{f}(\xi)\bar{f}(\eta)\}[1 - e^{-v\xi^2 t/\rho}][1 - e^{-v\eta^2 t/\rho}]\, d\xi\, d\eta, \quad (6.17a)$$

$$E\{v(r_1, t)v(r_2, t)\} = \int_0^\infty \int_0^\infty \xi\eta J_0(\xi r_1)J_0(\eta r_2) \cdot$$
$$E\{\bar{f}(\xi)\bar{f}(\eta)\}e^{-v\xi^2 t/\rho} e^{-v\eta^2 t/\rho} \, d\xi \, d\eta, \tag{6.19}$$

$$E\{\sigma(r_1, t)\sigma(r_2, t)\} = v^2 \int_0^\infty \int_0^\infty \xi^2\eta^2 J_1(r_1\xi)J_1(r_2\eta)e^{-v\eta^2 t/\rho} \cdot$$
$$E\{\bar{f}(\xi)\bar{f}(\eta)\}e^{-v\xi^2 t/\rho} \, d\xi \, d\eta. \tag{6.20}$$

If $E\{f(r)\} = v(r)$,

$$E\{f(r_1)f(r_2)\} = \phi(r_1, r_2); \tag{6.21}$$

then we have

$$E\{\bar{f}(\xi)\} = \int_0^\infty v(r)rJ_0(\xi r) \, dr,$$

$$E\{\bar{f}(\xi)\bar{f}(\eta)\} = \int_0^\infty \int_0^\infty \phi(r_1, r_2)r_1 r_2 J_0(\xi r_1)J_0(\eta r_2) \, dr_1 \, dr_2. \tag{6.22}$$

Substituting these values of $E\{\bar{f}(\xi)\}$ and $E\{\bar{f}(\xi)\bar{f}(\eta)\}$ in the questions above, we can get the statistical characteristics of w, v, and σ. For any specific set of values of r_1, r_2, and t the integrals involved can be computed numerically if the form of $v(r)$ and $\phi(r_1, r_2)$ is known.

6.2. Numerical Results

In this section we assume special forms for $v(r)$ and $\phi(r_1, r_2)$ and evaluate the integrals on the right-hand side of (6.18) through (6.20). Two different forms of $v(r)$ have been proposed by Subramanian and Kumaraswamy [17].

Model I. In this model, $v(r)$ is given by $v(r) = v_0 \exp(-kr/R)$. The maximum value of $v(r)$ is obviously at $r = 0$; it decreases as r increases. This is a reasonably good model, as the velocity would be greatest at the point of impact and would fall off at distances farther from that point.

Model II. Here $v(r)$ is given by

$$v(r) = \frac{v_0 kR}{2(1 - e^{-k})r} \exp\left(\frac{-kr}{R}\right).$$

We note that $v(r)$ is infinite at $r = 0$ and falls off to zero as r increases. This would serve as a better model, as it is only reasonable to expect the velocity at the point of impact to be very large and to be negligible at distances far removed from the point of impact. The constant factor is so chosen as to yield a mean velocity v_0 over a circle of radius R. The following form for $\phi(r_1, r_2)$ has also been assumed [17]:

$$\phi(r_1, r_2) = \frac{v_0{}^2 k^3 R}{4[2 - e^{-k}(k^2 + 2k + 2)]} e^{-k r_1/R} \delta(r_1 - r_2). \qquad (6.23)$$

The motivation for proposing the form in (6.23) is the reasonable assumption that the velocities at two points r_1 and r_2 separated by a large distance will not be correlated and that the correlation of velocities at two contiguous points r_1 and r_2 will be a monotonic decreasing function of r_1. The other factors are adjusted to satisfy the relation

$$4\pi^2 \int_0^R \int_0^R \phi(r_1, r_2) r_1 r_2 \, dr_1 \, dr_2 = \pi^2 R^4 V_0{}^2. \qquad (6.24)$$

The integral on the left-hand side is a measure of the average value of the square of the momentum over a circle of radius R.

In order to make a comparison with the results of [16], we make use of dimensionless variables given by the following scheme:

$$x = \frac{r}{R}, \qquad \tau = \frac{vt}{\rho R^2}, \qquad \Sigma = \frac{R\sigma}{vv_0},$$

$$V = \frac{2v}{v_0}, \qquad W = \frac{vw}{v_0 R^2 \rho}, \qquad \alpha = \xi R, \qquad \beta = \eta R, \qquad (6.25)$$

where v_0 is some characteristic velocity. The integrals in (6.14a) through (6.16a) have been evaluated numerically by Subramanian and Kumaraswamy [17] for various values of k and specified values of τ. Figures 6.1 to 6.6 give the mean values of the displacement, velocity, and stress corresponding to a set of values of k and τ. It is found that the graph of the mean value of the displacement corresponding to $k = 3$ in Model I agrees very favorably with the experimental curves given in [16].

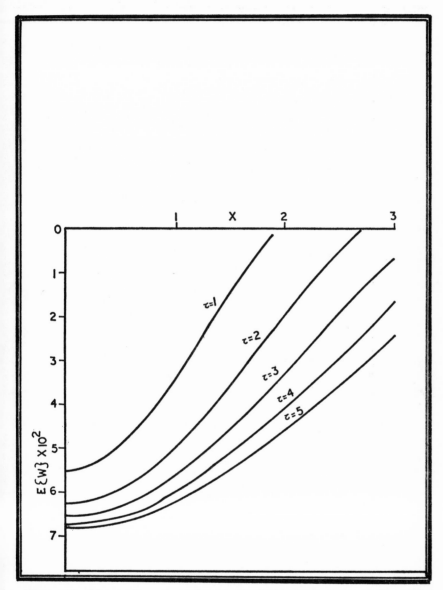

Figure 6.1. Mean value of nondimensional displacement versus nondimensional distance (Model I; $k = 3$).

9

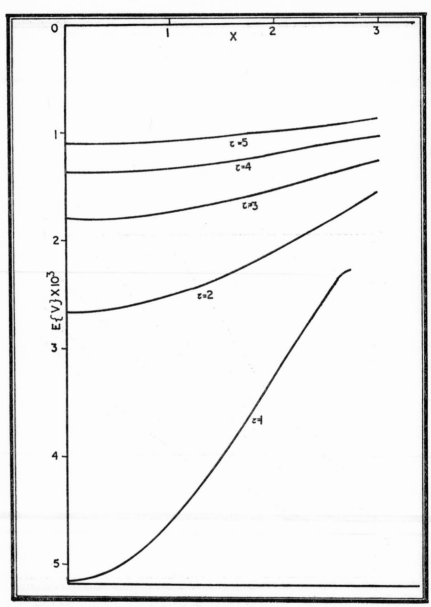

Figure 6.2. Mean value of nondimensional velocity versus nondimensional distance (Model I; $k = 3$).

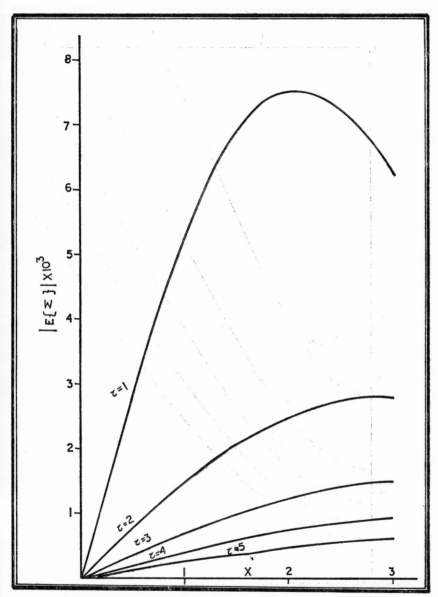

Figure 6.3. Mean value of nondimensional stress versus nondimensional distance (Model I; $k = 3$).

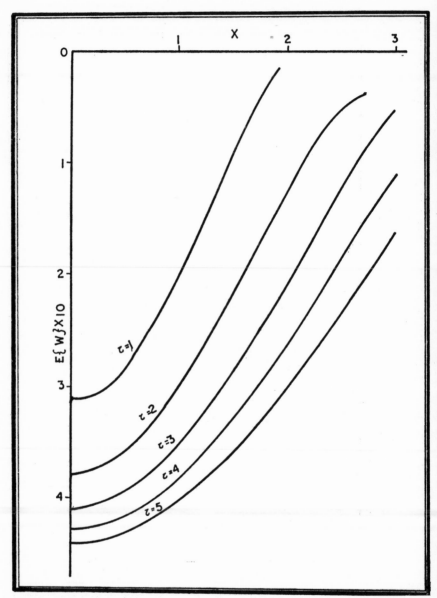

Figure 6.4. Mean value of nondimensional displacement versus nondimensional distance (Model II; $k = 1$).

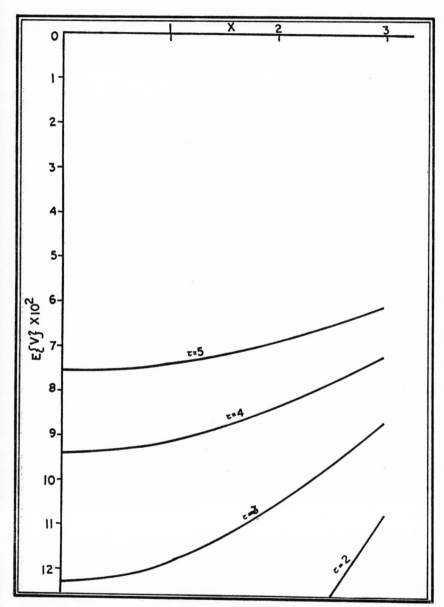

Figure 6.5 Mean value of nondimensional velocity versus nondimensional distance (Model II; $k = 1$).

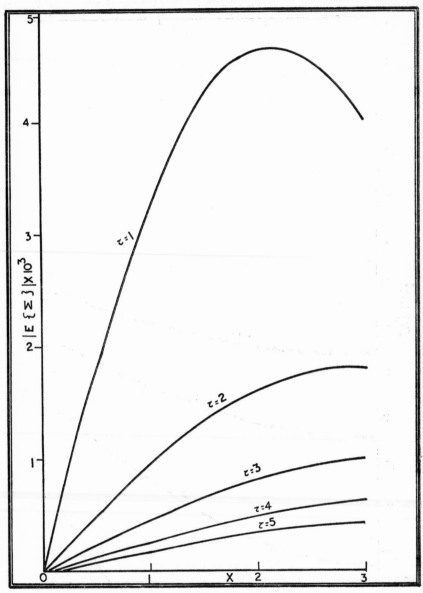

Figure 6.6. Mean value of nondimensional stress versus nondimensional distance (Model II; $k = 1$).

REFERENCES

1. M. J. Beran, *Statistical Continuum Theories*, Interscience, New York, 1968.
2. Y. K. Lin, *Probabilistic Theory of Structural Dynamics*, McGraw-Hill, New York, 1967.
3. T. K. Caughey, *J. Appl. Mech.*, **26**, *Trans. ASME*, **81**(1959), 341.
4. A. C. Eringen, *J. Appl. Mech.*, **24**, *Trans. ASME*, **79**(1957), 46.
5. W. T. Thompson and M. V. Barton, *J. Appl. Mech.*, **24**, *Trans. ASME* **79**(1957), 248.
6. S. H. Crandall and A. Yieldiz, *J. Appl. Mech.*, **29**, *Trans. ASME*, **84**(1962), 267.
7. R. Subramanian and S. Kumaraswamy, *J. Math. Phys. Sci.*, **3**(1969), 285.
8. S. H. Crandall and W. D. Mark, *Random Vibration in Mechanical Systems*, Academic Press, New York, 1963.
9. R. E. Herbert, *J. Acoust. Soc. Amer.*, **36**(1964), 2090.
10. R. E. Herbert, *Int. J. Solids Structures*, **1**(1965), 235.
11. W. E. Boyce, The Bending of Columns under Axial Loads of Random Eccentricity, Rensselaer Polytechnic Institute, Math. Report No. 43 (1961).
12. W. E. Boyce, *J. Aerospace Sci.*, **28**(1961), 307.
13. G. N. Lance, *Quart. Appl. Math.*, **15**(1956), 312.
14. S. K. Srinivasan, R. Subramanian, and S. Kumaraswamy, *Arch. Mechaniki Stosowanej*, **21**(1969), 191.
15. Alladi Ramakrishnan, *Proc. Cambridge Philos. Soc.*, **46**(1950), 595.
16. A. Pytel and N. Davids, *J. Franklin Inst.*, **276**(1963), 394.
17. R. Subramanian and S. Kumaraswamy, *Rozprawy Inzynierskie* (1970), (to be published).

Chapter 7

TRANSPORT PHENOMENA

1. INTRODUCTION

Many physical processes involve transport of mass or energy from one spatial point to another through a medium, which determines the nature of the transport in an essential manner. In the late nineteenth century, interest was centred on the study of diffusion of light by the atmosphere, and investigations of the elementary type of Boltzmann equations were carried out. The solution of the Milne problem relating to passage of radiation in a semi-infinite plane parallel gaseous atmosphere by Wiener and Hopf as early as 1931 stimulated an increase in interest in this field of study. However, the discovery of the neutron and the extensive efforts taken to use this particle as a missile to disrupt the nucleus and release vast stores of nuclear energy useful in both war and peace initiated a tidal wave of progress in these investigations. Neutron diffusion is now a separate branch of specialization which has led to the innovation of new computational techniques and novel types of mathematical analysis.

Even during the war, Ambartzumian [1] introduced an ingenious approach based on functional equations arrived at by intuitive principles of invariance in the field of radiative transfer. The older methods of using a linearized Boltzmann equation are treated by Chandrasekhar [2] and Bursbridge [3]. The ideas of invariance due to Ambartzumian were also handled by Chandrasekhar in his monograph. They were generalized and extended and built into a systematic tool to attack different problems in several fields by Bellman and Kalaba [4] and were applied to the study of reflection radiation from inhomogeneous space of different geometries. This systematic use of invariance concepts to arrive at functional equations is called the invariant imbedding method. Of course, they are related to procedures such as regeneration point techniques and backward equations in the generalized theory of branching processes dealt with by Harris [5], Ramakrishnan [6], and Bellman and Harris [7]. Bellman, Kalaba and Wing, in a series of papers, applied these methods (see Wing [8]) to neutron transport theory and showed the possibility of reducing two-point boundary value problems to initial value problems which are more amenable to efficient computational solutions.

The events occurring in the medium when neutrons or radiation passes through it are governed by probabilities. The results of interaction of the particles with the medium are again probabilistic in nature. Thus the transport theory of necessity can only be a probabilistic theory. The first half of this chapter deals with stochastic equations governing transport of particles like neutrons in reactors or the photons in radiative transfer through planetary atmospheres. However, transfer of energy from one spatial point to another is caused by the mechanism of wave motion also. The second half of the chapter is devoted therefore to the study of wave propagation in a stochastic medium. Many physical problems, such as scattering of sound waves by turbulent fluids, twinkling of the stars caused by the passage of their light through the atmosphere with refractive properties which are random, are some of the most common examples. A variety of methods are used in tackling the stochastic equations governing transport of waves (see Keller [9]).

Wave motion in random media can be divided broadly into two parts, one dealing with propagation in continuous random medium and the other describing the multiple scattering of the waves by randomly distributed scatterers in the medium. In spite of many difficulties, both physical and mathematical, in treating these processes described by linear random equations, very effective perturbative methods and related diagram techniques have been developed. Considerable progress has been made with one-dimensional models (see review by Frisch [10]; also see Kraichnan [11] and Tatarski [12]).

Wave propagation in random media is also closely related to many other fundamental problems in both mathematical and applied physics. There is a close analogy between this study and fundamental physical theories such as quantum field theory, equilibrium and nonequilibrium statistical mechanics, turbulence theory, and other related phenomena. Of course, the analogies are not accidental, and hence the introduction of many body techniques, such as diagrams and formal expansions, into the studies of random wave propagation are motivated by these analogies. It is hoped that further progress in the investigations of linear random equations will enrich these general theories by feedback.

In the following sections the emphasis will be on the various kinds of techniques developed to solve the different types of linear stochastic equations met with in various contexts of the transport processes. In Section 2 we begin the description of neutron multiplication processes.

2. NEUTRON TRANSPORT

Studies of neutron transport have become an integral part of nuclear reactor physics (see for example Davison [13]); such reactors are being built in increasing numbers to tap the energies stored in the nucleus. If a neutron beam is allowed to pass through a slab of target material, it will emerge with reduced intensity, owing to a variety of processes that take place during its passage through the material such as scattering absorption with accompanying γ emission or absorption leading to subsequent fission. The attenuation of the neutron beam due to various causes can be described by introducing the absorption cross section σ, which defines the probability of the neutron being absorbed for several reasons, and σ itself is the sum of these cross sections for these processes:

$\sigma = \sigma$ (elastic scattering) $+ \sigma$ (inelastic scattering) $+$

σ (fission) $+ \sigma$ (capture).

There is a quantum mechanical basis for the calculation of these cross sections. Also they can be deduced experimentally from various nuclear reactions. The cross sections depend on the medium through which the neutron diffuses and also on the energy of the neutron. According to their energies, the neutrons are classified as fast neutrons, intermediate neutrons, epithermal and thermal neutrons. The thermal neutrons have energies in the range 1 eV or less. Fast neutrons in a reactor are used as missiles to disrupt the nucleus and cause fission. They produce a copious supply of fast neutrons, called moderators which have to dissipate their high initial energies, with numerous collisions in the material incorporated in the reactors. The purpose of reducing the energy of neutrons, called thermalization of the neutrons, is to increase the fission cross section, thus securing and maintaining the chain process. The vast number of these neutrons present at any point in the reactor with varied energies and the rapid changes in their fluxes and densities make it possible to adopt the statistical approach in the study of the physical processes involved here and to employ the techniques of kinetic theory. Once these neutrons are produced, they are scattered and absorbed or lost as they move in the reactor. The number of particles keeps changing, and the central feature of these reactors is the so-called chain reaction by which we understand that the reactor mechanism is a self-sustaining process. This means that neutrons produced in fission are either lost to the medium or used up to make more neutrons come out in further fissioning processes, thus sustaining the chain reactions in the material. To follow up this complex history of these groups of neutrons from the viewpoint of stochastic theory and arrive at the criteria for the reactor to go critical is the main objective of neutron transport theory. These are very well described by Weinberg and Wigner [14]. First we describe the thermalization process of a single neutron before taking up the transport of the assembly of neutrons.

2.1. Lethargy Equations

In the thermalization process of a neutron we speak of the lethargy of the neutron. The stochastic process associated with the lethargy of the neutron was studied by Takacs [15]. If a neutron starts out with initial energy E_0 at time $t = 0$ and if its energy after many collisions degrades to $E(t)$, the lethargy of the neutron is defined as

$$L(t) = \log \frac{E_0}{E(t)} \quad \text{or} \quad E(t) = E_0 e^{-L(t)}. \tag{2.1}$$

We assume that, when a neutron traverses the medium:

(i) it suffers a collision with a nucleus in a distance Δt with probability $P_i(l)e^{-1/2} \Delta t$; and

(ii) the collision is a lethargy changing collision effecting a change $(l' - l)$ in its lethargy with probability $P_i(l)e^{-1/2}H_i(l, l') \Delta t \, dl'$.

$H_i(l, l')$ depends on the nature of the nucleii in the medium and is known to be a function of $(l - l')$ only, the function being zero for negative values of its argument. Defining $F(l, t)$ as the probability that the neutron has a lethargy equal to or less than l, by analyzing the possibilities in the interval t to $t + \Delta t$ we write

$$\frac{\partial F(l, t)}{\partial t} = - \int_0^l e^{-y/2} \left[\sum_{i=1}^{N} P_i(y) - \sum_{i=1}^{N} P_i(y)H_i(l - y) \right] d_y F(y, t) \tag{2.2}$$

with the initial conditions

$$F(l, 0) = 0, \quad l \geqslant 0,$$
$$= 1, \quad l < 0. \tag{2.3}$$

In following a single neutron, other interesting questions arise: What is the expected time required by the neutron to reach a given lethargy value? What is the mean number of collisions a neutron will undergo before reaching a certain lethargy value? These questions can be studied by defining suitable probability functions and forming the relevant differential or integro-differential equation.

2.2. Classical Boltzmann Equations

Apart from the thermalization problems with neutrons referred to in Section 2.1, the more important aspect is the actual density and flux of neutrons at every point in the medium, taking into account all the processes taking place in the medium. This problem is identical with the problems of radiative transfer or diffusion of light through the atmosphere. All these problems are described by one common type of transport equation, namely, the Boltzmann equation. Methods of solution of these equations have been developed beginning from the solution of Milne problem by Wiener and Hopf (see Hopf [16]). We now deal with the classical approach in some detail before going on to the imbedding methods referred to in the introduction.

The neutron is considered a point particle and it suffers collisions with the nuclei in the medium and either can be scattered in different directions with different velocities or may be absorbed. A collision may result in a fission process which gives rise to other neutrons. There may also exist other sources of neutrons in the medium. We assume that the number of neutrons available is so large that they can be considered to form a statistical assembly, and therefore we can talk of the mean number $N(\mathbf{r}, v, \mathbf{\Omega}, t) \, dv \, d\mathbf{\Omega}$ of neutrons per unit volume in the velocity range $(v, v + dv)$ lying in the cone around $\mathbf{\Omega}$ at time t at the point $\mathbf{r}$. Thus this is a product density in the product space of v and $\mathbf{r}$. Another important assumption is that the neutron fluxes do not affect the nature of the medium of the reactor. Also we neglect neutron-neutron collisions. These are the essential reasons why the Boltzmann equation in this situation, unlike that in the kinetic theory, is *linear*. Let α be the inverse of the total neutron free path l and let $\beta = \alpha c$ be the mean number of secondaries produced per unit path, c being the mean number of secondary neutrons produced in a collision. Then we can write the transport equation as

$$\frac{\partial N(\mathbf{r}, v, \mathbf{\Omega}, t)}{\partial t} + v\mathbf{\Omega} \cdot \text{grad } N(\mathbf{r}, v, \mathbf{\Omega}, t) + v\alpha N(\mathbf{r}, v, \mathbf{\Omega}, t) =$$

$$\iiint \beta v' N(\mathbf{r}, v', \mathbf{\Omega}', t) f(v'\mathbf{\Omega}' \to v\mathbf{\Omega}) \, dv' \, d\mathbf{\Omega}' + S(\mathbf{r}, v, \mathbf{\Omega}, t), \quad (2.4)$$

where $cf(v'\mathbf{\Omega} \to v\mathbf{\Omega}) \, dv \, d\mathbf{\Omega}$ is the mean number of neutrons produced in the velocity range dv and $d\mathbf{\Omega}$ by the collision of a neutron of velocity v' and angle $\mathbf{\Omega}'$ with a nucleus of the medium, and S is the source function. The quantity f is normalized to unity. Let us make simplifying assumptions that only one velocity is allowed and that the scattering is isotropic. Also, if there is plane symmetry in the problem, $f = 1$ and N depends only on the position coordinate x, and the direction coordinate Ωx becomes μ, the cosine of the angle of the direction with the x-axis. Equation (2.4) then reduces to

$$\frac{\partial N(x, \mu, t)}{\partial t} + \mu v \frac{\partial N(x, \mu, t)}{\partial x} + v\alpha N(x, \mu, t) =$$

$$\frac{c}{2}v\alpha \int_{-1}^{1} N(x, \mu', t)\, d\mu' + S(x, \mu, t). \qquad (2.5)$$

The solutions of this equation treated as a characteristic value problem were studied by Case [17]. But the solution of this equation brings up many mathematical problems, details of which are found in the review by Wigner [18]. Study of critical size of the medium and the connected problems are also reviewed by Tait [19]. However, to understand the phenomenon of the reactor going critical, we take a simple one-dimensional rod and write the neutron flux equations inside the rod. We follow the classical approach to the transport problem and obtain solutions of the kinetic equations which indicate explicitly how the critical length of the rod can be found as a function of the cross sections for the occurrence of several processes in the medium.

Following the procedure adopted by Ramakrishnan, Vasudevan, and Rajagopal [20], we take the case in which all the neutrons have the same magnitude of velocity, $|v|$. They undergo absorption at any point x in the medium with probability $c\,dx$ and suffer fission with a probability $a\,dx$. During each collision the original neutron is replaced by two neutrons, one going with velocity $+ v$ to the right and another with velocity $- v$ to the left. The quantity of interest is the mean number of neutrons that emerge at both ends of the rod at $x = 0$ and $x = L$. To this end, we define the product densities $f^R(x, t)$ and $f^L(x, t)$ corresponding to the particles that cross x to the right and left respectively at time t. Coupled differential equations can be written for f^L and f^R by considering what happens at x between t and $t + \Delta$. We have

$$\frac{\partial f^R(x, t)}{\partial t} = af^L(x, t) - cf^R(x, t) - \frac{1}{v}\frac{\partial f^R(x, t)}{\partial t},$$

$$\frac{\partial f^L(x, t)}{\partial t} = - af^R(x, t) + cf^L(x, t) + \frac{1}{v}\frac{\partial f^L(x, t)}{\partial t}. \qquad (2.6)$$

We can take the Laplace transform with respect to t to obtain

$$\frac{\partial}{\partial x}\begin{pmatrix} P^R(x, s) \\ P^L(x, s) \end{pmatrix} = \begin{pmatrix} - c - s/v & a \\ - a & c + s/v \end{pmatrix}\begin{pmatrix} P^R(x, s) \\ P^L(x, s) \end{pmatrix} + \begin{pmatrix} 1/v & f^R(x, 0) \\ - 1/v & f^L(x, 0) \end{pmatrix},$$

where P^R and P^L are the Laplace transforms of f^R and f^L and s is the transform variable. If s is put equal to 0 in the equations above, the functions $P^R(x, 0)$

and $P^L(x, 0)$ straightaway yield the total number of neutrons at x moving to the right or left for all time.

Using the initial conditions corresponding to the situation that the process is triggered by a single neutron at $x = 0$, $t = 0$, we arrive at the mean number that emerge at the two ends for all time:

$$P^L(0, 0) = \frac{a \sin mL}{c \sin mL + m \cos mL},$$

$$P^R(0, L) = \frac{m}{c \sin mL + m \cos mL}$$

with $m = (a^2 - c^2)^{\frac{1}{2}}$. If $c = 0$, we have

$$P^L(0, 0) = \tan aL = M,$$

$$P^R(L, 0) = \sec aL = N. \tag{2.9}$$

We observe that the process becomes critical for that length of the rod which makes $P^R(L, 0)$ infinite. In general, taking c into account, the critical length L_c is given by

$$\cot((a^2 - c^2)^{\frac{1}{2}} L_c) = \frac{-c}{a^2 - c^2}. \tag{2.10}$$

If we do not put s equal zero in (2.7) and solve for $P^L(0, s)$ and $P^R(L, s)$, we obtain

$$P^L(0, s) = \frac{av \sinh mL}{s \sinh mL + vm \cosh mL},$$

$$P^R(L, s) = \frac{vm}{s \sinh mL + vm \cosh mL}, \tag{2.11}$$

with $m = (v^2/s^2 - a^2)^{\frac{1}{2}}$. Inverting the expressions for $P^L(0, s)$ and $P^R(L, s)$, we can obtain $f^R(L, \tau)$ and $f^L(0, \tau)$ which, when integrated over τ from 0 to t, give $N(t)$ and $M(t)$, the mean numbers that emerge at the ends $x = L$ and $x = 0$ for a time t.

If, however, at each collision of a neutron two pairs of particles are produced, one pair moving left and the other moving right always with the same velocity, it is necessary to introduce multiple product densities (first introduced by Ramakrishnan and Srinivasan [21]) in the context of age distribution in a population with twins and multiplets. The situation in which there are collisions producing two neutrons each moving in opposite directions and collisions in which there are two pairs of neutrons, the pairs going in opposite directions and each neutron suffering backscattering also with certain probabilities are considered in [20]. Four coupled differential equations for the product densities $f^L(x, t)$, $f^R(x, t)$ defined earlier, and $f^{2L}(x, t)$ and $f^{2R}(x, t)$ representing the mean number of pairs crossing x to the left or the right are written. Transform techniques are used to solve these equations and obtain the mean numbers that emerge at either end for all time. Thus all these equations for the internal fluxes are two-point boundary value problems. In Section 3 we convert them into initial value problems by the use of the imbedding technique.

3. INVARIANT IMBEDDING APPROACH

Several one-dimensional problems in neutron multiplication can be successfully handled by the invariant imbedding approach developed in a series of papers by Bellman, Kalaba, and Wing [22]. The great advantage of this method is that these problems of transport theory can be transformed from eigenvalue problems into initial value problems well suited to high-speed digital computation. The philosophy of this approach is to imbed a particular process within an appropriate class of processes and then derive the relationships existing between them. We arrive at functional equations which enable us either to obtain new analytic results or to forge new computational tools. We describe here some of the one-dimensional neutron multiplication problems treated in this fashion.

3.1. Criticality

If a neutron impinges at one end of the one-dimensional rod of fissionable material and if $a\Delta$ is the probability of fission in a small distance Δ, we are interested in the expected number of neutron fluxes reflected back at the same end or transmitted through the rod emerging at the other end. In the imbedding method we relate the reflected or transmitted flux for a rod of length x with the corresponding fluxes for rods of different lengths. To do this we define the reflection function $r(x)$ as the mean number reflected over all time from a rod of length x and resulting from a trigger neutron incident at time zero. For a rod of length $x + \Delta$ we can analyze the possible outcome at the first infinitesimal interval Δ after the entry of the neutron. A fission may take place giving two neutrons, one going to the right which emerges as reflected flux while the other, moving left, is the incident flux at x for a rod of length x. If there is no fission in Δ, then the original neutron itself will be incident at x and will cause a reflected flux $r(x)$. Some of the reflected neutrons coming out at x may again undergo fission, producing neutrons some of which move to the left and others to the right. Although the physical process and the mathematical counterpart are exceedingly complex if account is taken of all reflections and fissions, this intricate bookkeeping is unnecessary because Δ

is infinitesimal and we can neglect all processes which are of order $o(\Delta)$. If we add up all the associated probabilities, we have

$$r(x + \Delta) = a\Delta[r(x) + 1] + [1 - a\Delta]r(x) + [1 - a\Delta]r(x)a\Delta r(x) + o(\Delta).$$
(3.1)

Proceeding to the limit as Δ tends to zero, we obtain the Riccati differential equation for $r(x)$:

$$r'(x) = a[1 + r^2(x)]$$
(3.2)

with $r(0) = 0$ as the initial condition. The solution is given by $r(x) = \tan ax$, which indicates that the critical length is $\pi/2a$, as obtained in the classical approach using the coupled equations for the internal fluxes.

For the transmitted flux $t(x)$ we have, by the same type of argument,

$$t(x + \Delta) = t(x) + r(x)a\Delta t(x) + o(\Delta).$$
(3.3)

Letting Δ tend to zero, we have

$$t'(x) = ar(x)t(x).$$
(3.4)

With initial condition $t(0) = 1$, we find the solution for $t(x)$ as

$$t(x) = \sec ax.$$
(3.5)

By considering the events in the rod at the far end, for the reflected flux we can obtain another equation,

$$r(x + \Delta) = r(x) + t(x)a\Delta t(x) + o(\Delta),$$
(3.6)

which implies that

$$r'(x) = at^2(x).$$
(3.7)

Comparing this with (3.2), we see that

$$1 + r^2(x) = t^2(x).$$
(3.8)

This relation between transmission and reflection coefficients is analogous to the Stokes relation in optics.

If we take a more realistic model, (3.2) will become more complicated. However, the critical length problem is an initial value problem, and so we can integrate (3.2) or its generalization step by step until we obtain a solution which blows up. In fact, we need not integrate till we reach the infinite solution. Counting successive reflections and transmission, we can show that

$$r(x + y) = r(x) + t(x)r(y)t(x) + t(x)r(y)r(x)r(y)t(x) + \cdots$$
$$= \frac{r(x) + r(y)}{1 - r(x)r(y)}.$$
(3.9)

Hence it follows that $r(x_c/2) = 1$ if x_c is the critical length, and so we need to follow up the solution to the point when $r(x)$ becomes unity for a complicated equation; this is the decided advantage in numerical computation.

In certain situations we may like to have a more detailed picture of the reflected and transmitted flux. Toward this end, we may define probabilities $P_n(x)$ representing the probability that n neutrons are reflected from a rod of length x over all time as a result of a trigger neutron at one end at time zero. As before, we take a rod of length $(x + \Delta)$. We may write the following relation, which is valid up to the first power in Δ, taking into account all the possibilities that can occur:

$$P_n(x + \Delta) = (1 - a\Delta)P_n(x)(1 - n\Delta a) +$$

$$\sum_{k=1}^{n} P_k(x)ka\Delta P_{n-k}(x) + a\Delta P_{n-1}(x) + o(\Delta). \qquad (3.10)$$

Passing to the limit as Δ tends to zero yields the following system of differential equations:

$$P_n'(x) = - (n + 1)aP_n(x) + aP_{n-1}(x) +$$

$$a\sum_{k=1}^{n} kP_k(x)P_{n-k}(x), \qquad n = 0, 1, 2, \ldots,$$

$$P_{-1}(x) = 0, \qquad (3.11)$$

with

$$P_n(0) = \begin{matrix} 1, & n = 0, \\ 0, & n \neq 0. \end{matrix} \qquad (3.12)$$

Computer calculations using the recursive nature of these relations are particularly simple since they are now initial value problems. The calculations indicate that the criticality will be obtained for a length $\pi/2a$. In a similar manner, the equation for the probability that n particles are transmitted can also be arrived at:

$$q_n'(x) = - anq_n(x) + a\sum_{k=1}^{n} kq_k(x)P_{n-k}(x), \qquad n = 1, 2, \ldots, \qquad (3.13)$$

with

$$q_n(0) = \begin{matrix} 1, & n = 0, \\ 0, & n > 0. \end{matrix} \qquad (3.14)$$

From (3.1) we can easily see that the expected number $r(x) = \sum_{n=0}^{\infty} nP_n(x)$ obeys the Riccati equation (3.2). If we are interested in the quantity $M(x) = \sum n(n - 1)P_n(x)$, we can easily obtain from (3.11) and (3.12)

$$M(x) = \tfrac{2}{3}\{\tan^3 ax + \sec^3 ax - 1\}. \qquad (3.15)$$

The standard derivation $\sigma(x)$ of the number of reflected particles is given by

$$\sigma(x) = [\tfrac{2}{3}(\tan^3 ax + \sec^3 ax - 1) + \tan ax - \tan^2 ax]^{\frac{1}{2}}. \qquad (3.16)$$

10

3.2. Time Dependence and Energy Dependence

To calculate the moments of the reflected and transmitted flux, we can use the product density technique (see Ramakrishnan [23]). Defining $u_1(x, t)$ and $u_2(x, t_1, t_2)$ as the product densities of degree 1 and 2 of reflected neutrons at edge x, the imbedding equations satisfied by these correlation functions can be written (see Bellman *et al*, [24]).

The functional equation satisfied by $u_1(x, t)$ is given by

$$u_1\left(x + \Delta, t + \frac{2\Delta}{v}\right) =$$

$$(1 - a\Delta)\left[u_1(x, t) + a\Delta \int_0^t u_1(x, s)u_1(x, t - s) \, ds\right] +$$

$$a\Delta[\delta(t) + u(x, t)] + o(\Delta), \qquad (3.17)$$

where v is the velocity of the neutron. When Δ tends to zero, we find

$$\frac{\partial u_1(x, t)}{\partial x} + \frac{2}{v}\frac{\partial u_1(x, t)}{\partial t} = a \int_0^t u_1(t, s)u_1(x, t - s) \, ds + a \, \delta(t). \qquad (3.18)$$

Defining the Laplace transform of $u_1(x, t)$ as $G_1(x, p)$, we obtain

$$\frac{\partial G_1(x, p)}{\partial x} + \frac{2p}{v}G_1(x, p) = a[(G_1(x, p))^2 + 1]. \qquad (3.19)$$

To obtain the flux for all time we put $p = 0$ and obtain the usual Riccati equation (3.2). In an analogous manner we can write the imbedding equation for $u_2(x, t_1, t_2)$, the Laplace transform of which obeys the equation

$$\frac{\partial G_2(x, p_1, p_2)}{\partial x} + 2\left(\frac{p_1}{v} + \frac{p_2}{v}\right)G_2(x, p_1, p_2) =$$

$$aG_2(x, p_1, p_2)[G_1(x, p_1) + G_1(x, p_2)] +$$

$$aG_1(x, p_1 + p_2)[G_2(x, p_1, p_2) + G_1(x, p_1) + G_1(x, p_2)] +$$

$$aG_1(x, p_2) + aG_1(x, p_1), \qquad (3.20)$$

where $G_2(x, p_1, p_2)$ is the Laplace transform of $u_2(x, t_1, t_2)$. The quantity $S(x)$, defined by

$$S(x) = \int_0^\infty \int_0^\infty u_2(x, t_1, t_2) \, dt_1 \, dt_2, \qquad (3.21)$$

satisfies the equation

$$\frac{d}{dx}S(x) = a[3S(x)r(x) + 2[r(x)]^2 + 2r(x)]. \qquad (3.22)$$

Solution of $S(x)$ will lead to the determination of the fluctuation in the number reflected over all time.

Since, in reactor theory, the energies of the neutron beams which are reflected and transmitted are also important, it is necessary to introduce the product densities defined over the energy space. Thus it is convenient to introduce the function $u_1(x, E, E_0)$, where $u_1(x, E, E_0)\, dE$ denotes the probability that a neutron has an energy between E and $E + dE$ after reflection from a rod of length x, E_0 being the energy of the trigger neutron. It is generally assumed that during fission the probability that a neutron of energy E yields a neutron moving in the same direction, its energy lying between E_1 and $E_1 + dE_1$, and another neutron moving in the opposite direction in the rod, its energy lying between E_2 and $E_2 + dE_2$, is given by $Q(E, E_1, E_2)\, dE_1\, dE_2$. If we assume that the probability of formation of fission by a neutron of energy E is $a(E)$ per unit length, we obtain

$$u_1(x, E, E_0) = [1 - a(E_0)\Delta]u_1(x - \Delta, E, E_0)\{1 - a(E)\Delta\} +$$

$$\Delta \int\int u_1(x - \Delta, E_1, E_0)a(E_1)Q(E_1, E, E_2)\, dE_1\, dE_2 +$$

$$\Delta \int\int\int u_1(x - \Delta, E_1, E_0)Q(E_1, E_2, E_3)a(E_1)u_1(x - \Delta, E, E_3)\, dE_1\, dE_2\, dE_3 +$$

$$a(E_0)\Delta[\int Q(E_0, E, E_1)\, dE_1 +$$

$$\int\int Q(E_0, E_1, E_2)u_1(x - \Delta, E, E_2)\, dE_1\, dE_2] + o(\Delta). \tag{3.23}$$

We can now pass on to the limit when Δ goes to zero and we obtain the relevant differential equations for u_1. Similarly, second-order energy-dependent correlation functions $u_2(x, E_1, E_2, E_0)$ can be defined for the reflected neutron flux and imbedding equations can be formed. A knowledge of u_1 will yield the average energy of the neutron flux, and fluctuations in the average energy released from the medium over all time can be computed by solving for u_1 and u_2 from these equations (see Bellman, Kalaba, and Vasudevan [25]).

Neutron transport problems in a slab and for media of different geometries with angle-dependent reflected and transmitted fluxes have been the subject matter of a number of papers. Transport of electrons in a cathode tube and the dielectric breakdown of an one-dimensional gas leading to the phenomenon of Townsend avalanche have also been investigated by means of the imbedding method (see Bellman et al. [25]). For a survey of the various applications of the imbedding method of arriving at the functional equations, the reader is referred to the very interesting and comprehensive article by Bellman et al. [22].

4. TRANSPORT IN ASTROPHYSICS

The mathematical problem of radiative transfer theory is concerned with the solution of the integro-differential equation connected with the specific intensity of radiation at any point in the medium. The photons which are the carriers of the radiant energy are absorbed or scattered, and a part of the absorbed energy is also re-emitted. The amount of energy transported across an element of area $d\sigma$ in a specified frequency interval $(v, v + dv)$ in a direction making an angle θ with the normal to $d\sigma$ within an elementary solid angle $d\Omega$ in time dt is given by

$$dE_v = I_v \cos \theta \, d\sigma \, d\Omega \, dv \, dt, \tag{4.1}$$

where I_v is the specific intensity. A pencil of radiation of intensity I_v passing through a medium of thickness ds in the direction of its propagation will suffer absorption, and the change in I_v is given by

$$dI_v = - K_v \rho I_v \, ds, \tag{4.2}$$

where ρ is the density of matter and K_v is the mass absorption coefficient for frequency v.

The emission coefficient j_v is defined in such a way that the radiant energy emitted by an element of mass dm in a solid angle $d\Omega$ in the frequency range $(v, v + dv)$ in time dt is $j_v \, dm \, d\Omega \, dv \, dt$. The ratio of emission coefficient to absorption coefficient is known as the source function:

$$J_v = \frac{j_v}{K_v}. \tag{4.3}$$

For a plane parallel atmosphere with multiple scattering of photons in semi-infinite space (Milne's problem) for diffuse radiation, it has been shown (see Chandrasekar [2], Bursbridge [3]) that $J_v(\mu, t)$ satisfies an integro-differential equation (μ is the cosine of the angle between the direction of propagation and the axis perpendicular to the planes at the boundary):

$$\frac{\partial J(\mu, t)}{\partial t} = \frac{- J(\mu, t)}{\mu} + \tfrac{1}{2} J(\mu, 0) \int_0^1 J(\mu', t) \frac{d\mu'}{\mu'}. \tag{4.4}$$

Ueno [26] showed that, in the case of a plane parallel atmosphere, $P(\mu, t)$, the probability frequency function of a light beam at depth t, with cosine angle given by μ to the normal is governed by an integro-differential equation similar to (4.4). If $P(\mu, \mu', t) \, d\mu$ is the probability of reaching the value μ from μ' when radiation goes through a depth t in the medium, we can write the Chapman-Kolmogorov equation:

$$P(\mu, \mu', t) = \int_0^1 P(\mu'', \mu', t - \tau) P(\mu, \mu'', \tau) \, d\mu'', \tag{4.5}$$

where τ is arbitrary. Let us assume that $P(\mu, \mu', \tau)$ for small values of τ takes the following form (see Section 3.1 of Chapter 1):

$$P(\mu, \mu', \tau) = R(\mu, \mu')\tau + \delta(\mu - \mu')[1 - \tau\int R(\mu'', \mu')\,d\mu''] + o(\tau). \tag{4.6}$$

In addition, if we assume that

$$R(\mu, \mu') = \frac{P(\mu, 0, 0)}{2\mu'} \tag{4.7}$$

and require that

$$\int R(\mu, \mu')\,d\mu = \frac{1}{\mu'}, \tag{4.8}$$

then we easily obtain from the Kolmogorov equation (4.5) the following integro-differential equation when we pass to the limit τ tending to zero:

$$\frac{\partial P(\mu, \mu', t)}{\partial t} = -\frac{P(\mu, \mu', t)}{\mu} + \tfrac{1}{2}P(\mu, 0, 0) \int_0^1 P(\mu'', \mu', t)\frac{d\mu''}{\mu''}. \tag{4.9}$$

Comparison of (4.4) and (4.9) implies the equivalence of $P(\mu, \mu', t)$ and $J(\mu, t)$.* The solutions of equations of radiative transfer have been worked out by several authors.

The same problem has been reformulated as an initial value problem by Bellman and Kalaba [27], who defined the reflected mean specific intensity $I(\psi, \theta, x)$ (in a direction θ per unit area on the face of a slab of thickness x) due to a beam of unit intensity impinging at an angle ψ. If the absorption coefficient per unit thickness is a and the albedo is λ, by the usual arguments of the imbedding method we obtain an integro-differential equation for the mean intensity I:

$$\frac{dI(\psi, \theta, x)}{dx} = \frac{a(x)\lambda(x)}{4\pi \cos\psi} - a(x)\left[\frac{1}{\sec\theta} + \frac{1}{\sec\psi}\right]I(\psi, \theta, x) +$$

$$\frac{a(x)\lambda(x)}{2\cos\psi}\int_0^{\pi/2} I(\psi', \theta, x)\sin\psi'\,d\psi' +$$

$$\frac{a(x)\lambda(x)}{2}\int_0^{\pi/2} I(\psi, \theta', x)\tan\theta'\,d\theta' +$$

$$\pi a(x)\lambda(x)\int_0^{\pi/2}\tan\theta'\, I(\psi, \theta', x)\,d\theta'\int_0^{\pi/2} I(\psi', \theta, x)\sin\psi'\,d\psi' \tag{4.10}$$

with the initial condition

$$I(\psi, \theta, 0) = 0. \tag{4.11}$$

* Ueno proved the equivalence of $P(\mu, \mu', t)$ and $J(\mu, t)$ by showing that the assumptions (4.7) and (4.8) are plausible for the Markovian nature of the multiple scattering processes responsible for the radiative transport.

A vast battery of methods including numerical procedures, transform techniques, and other approximations have been constructed to solve these equations which, though nonlinear, are initial value problems. For detailed discussion of the importance of applying invariance principles, the reader is referred to the volume by Bellman, Kalaba, and Prestrud [28] in this series.

5. STOCHASTIC WAVE PROPAGATION

One of the familiar methods of transport of energy is through wave motion. Since the square of the amplitude of a wave, $\psi(x, t)$, at any space-time point is the energy at that point, propagation of a wave through a medium transmits energy from one region to another. This transmission is characterized by the physical properties of the medium such as the refractive index $n(x, t)$ of the medium. The physical laws governing wave motion imply that $\psi(x, t)$, the amplitude, satisfies a linear differential equation of the hyperbolic type. A function of the quantity $n(x, t)$ occurs as a coefficient in this equation. If the characteristics of the medium are inhomogeneous and stochastic in nature, we encounter a random wave equation. We may also have a random medium which may consist of a random distribution of discrete scatterers. This description is appropriate when we think of the phenomenon of the propagating wave on a molecular scale. Here also the process can be described by a stochastic equation.

Wave propagation in continuous random media occurs in the case of the scattering of radiowaves by tropospheric turbulence, twinkling of the stars, and in many other contexts. Multiple scattering of waves in media with random scatterers are met with in the phenomenon of molecular scattering of light, processes responsible for the dielectric constant of nonpolar gases; motion of an electron in a lattice with random impurities. Since the theory of partial differential equations with nonconstant coefficients is already difficult, the solutions of equations with random coefficients can be more so, since the solution of the linear equation may depend nonlinearly on the stochastic quantities. However, considerable progress has been made, at least with one-dimensional stochastic propagation. For the case of the continuous medium, a number of workers have contributed to the growth of the subject (see, for example, Keller [29]). Perturbation methods as well as functional techniques have been put to use in solving the stochastic equations. In Section 5.1 we deal with a continuous random medium.

5.1. Continuous Random Medium

There are usually two types of methods for solving these stochastic equations. The straightforward method is to obtain the solution of the equations explicitly either completely or by an iterative procedure. Then we compute the mean and moments using the probabilistic structure of the quantities involved in the solution. An alternative procedure utilizes the randomness of the system to write the equations governing the correlations of the solution $\psi(x, t)$. The averaged equations involve correlations between the solution ψ and other stochastic coefficients occurring in the equation. In this way each moment equation may be expressed in terms of successively higher-order moments. However, in some of these methods the hierarchy is broken off under some assumptions expressing moments of a particular order in terms of lower-order moments, since otherwise the problem becomes unsolvable. Some significant results in the theory of propagation in random media have been obtained using such methods. A clear understanding of the reason why such methods are successful may be rewarding in many other situations.

Like sound waves and electromagnetic waves, the waves which propagate in a random medium may be of different types. We shall assume that non-linear effects are neglected. Linear waves are governed by partial differential equations of the hyperbolic type, the coefficients characterizing the medium. The prototype of many wave equations is

$$\frac{\partial \psi}{\partial t} = L_0 \psi + L\psi, \tag{5.1}$$

where ψ is the scalar or vector-valued function of x and t is the wave function, L_0 is the nonrandom linear partial differential operator, usually with constant coefficients, and L contains coefficients which are random functions with mean values equal to zero. For harmonic time-dependent waves, we may, instead of with (5.1), work with the equation

$$(L_0 + L)\psi = g, \tag{5.2}$$

where g is the source term. As a typical case, radiation of scalar waves by harmonic point source in a lossless, homogeneous isotropic time-independent random medium, we can take the Helmholtz equation

$$\Delta\psi(\mathbf{r}) + k_0^2 n^2(\mathbf{r})\psi(\mathbf{r}) = \delta(\mathbf{r}), \tag{5.3}$$

where k_0 is the free space wave number and n is the refractive index, which is random and can be expressed by

$$n^2(\mathbf{r}) = 1 + \mu(\mathbf{r}) \tag{5.4}$$

(μ is a random function with zero mean and finite moments). To ensure the existence of the solution, proper conditions like the Sommerfeld radiation condition can be assumed to exist. If $G^0(\mathbf{r}, \mathbf{r}')$ is the free space Green's function of this problem, we can write the solution to (5.3) as

$$\psi(\mathbf{r}) = \int G^0(\mathbf{r}, \mathbf{r}')\psi(\mathbf{r}')d^3r' - k_0^2\int G^0(\mathbf{r}, \mathbf{r}')\mu(\mathbf{r}')\psi(\mathbf{r}')\, d^3r'. \tag{5.5}$$

This integral formulation of the problem is better suited for probabilistic situations, since the existence of derivatives of random functions is always a moot question. Alternatively, the solution of the above can also be related to the full Green's function $G(\mathbf{r}, \mathbf{r}')$ given by

$$G(\mathbf{r}, \mathbf{r}') = G^0(\mathbf{r}, \mathbf{r}') - k_0{}^2 \int G_0(\mathbf{r}, \mathbf{r}_1)\mu(\mathbf{r}_1)G(\mathbf{r}_1, \mathbf{r}')\, d^3r_1. \tag{5.6}$$

This is advantageous if we have a continuous density of sources in the medium, since Green's function describes the effect of a unit source at the origin. The solution for such a case is easily obtained as

$$\psi(\mathbf{r}) = \int G(\mathbf{r}, \mathbf{r}')g(\mathbf{r}')\, d^3r'. \tag{5.7}$$

In many physical and technical problems we are interested in the calculation of the mean of $G(\mathbf{r}, \mathbf{r}')$ or covariance of the Green's functions. The aim is to calculate the mean amplitude $\mathbf{E}\{\psi\}$ or mean intensity $\mathbf{E}\{|\psi(r)|^2\}$ at each point of the field.

If the fluctuation in the random parameter is small, we can replace L in (5.1) by $\varepsilon L_1 + \varepsilon^2 L_2$, where ε is a small parameter, and seek an expansion of ψ or G in a power series in the parameter ε. However, we should be aware of the difficulty inherent in a perturbation expansion. We may meet with the secular term "nuisance" if we stop with a finite order in the series in the averaged perturbation series, which therefore cannot be used asymptotically. These problems have been studied by a number of authors (see, for example, Kraichnan [30] and Richardson [31]).

Including the $\partial/\partial t$ term in the operator L_0 in (5.1), we write

$$L\psi = (L_0 + \varepsilon L_1 + \varepsilon^2 L_2)\psi = g. \tag{5.8}$$

If ψ^0 is the solution of

$$L_0\psi^0 = g, \tag{5.9}$$

assuming $L_0{}^{-1}$ exists, we can use an iteration procedure which, after a number of manipulations, leads to an equation for the expected value of ψ given by

$$\begin{aligned}\mathbf{E}\{\psi\} = \psi^0 &- \varepsilon L_0{}^{-1}\mathbf{E}\{L_1\}\mathbf{E}\{\psi\} + \\ &\varepsilon^2 L_0{}^{-1}[\mathbf{E}\{L_1 L_0{}^{-1} L_1\} - \mathbf{E}\{L_1 L_0{}^{-1}\}\mathbf{E}\{L_1\} - \\ &\mathbf{E}\{L_2\}]\mathbf{E}\{\psi\} + O(\varepsilon^2).\end{aligned} \tag{5.10}$$

Assuming that $L_2 = 0$ and $\mathbf{E}\{L_1\} = 0$ and operating $L_0 + L_1$ on both sides of (5.10), we arrive at

$$[L_0 - \varepsilon^2 \mathbf{E}\{L_1 L_0{}^{-1} L_1\}]\mathbf{E}\{\psi\} = g. \tag{5.11}$$

If we define the inverse of L_0 by

$$L_0{}^{-1}f(\mathbf{r}) = \int G^0(\mathbf{r}, \mathbf{r}')f(\mathbf{r}')\, d^3r', \tag{5.12}$$

we can rewrite (5.11) as

$$L_0(\mathbf{r})\mathbf{E}\{\psi(\mathbf{r})\} - \varepsilon^2 \int \mathbf{E}\{L_1(\mathbf{r})G^0(\mathbf{r}, \mathbf{r}')L_1(\mathbf{r}')\}\mathbf{E}\{\psi(\mathbf{r}')\} \, d^3r' = g(\mathbf{r}). \quad (5.13)$$

Applying the theory above to a scalar wave where the operators L_0, L_1 are given by

$$L_{0'} = \Delta - \frac{1}{c^2}\frac{\partial^2}{\partial t^2},$$

$$L_1 = -\frac{2\mu}{c^2}\frac{\partial^2}{\partial t^2} \quad (5.14)$$

and assuming

$$\mathbf{E}\{\mu(\mathbf{r}, t)\} = 0, \quad (5.15)$$

we find that, for free space ($g = 0$), (5.13) reduces to

$$\left(\Delta - \frac{1}{c^2}\frac{\partial^2}{\partial t^2}\right)\mathbf{E}\{\psi(\mathbf{r}, t)\} + \frac{\varepsilon^2}{\pi c^4}\int\frac{R(|\mathbf{r} - \mathbf{r}'|)}{|\mathbf{r} - \mathbf{r}'|}\frac{\partial^4}{\partial t^4} \cdot$$

$$\mathbf{E}\left\{\psi\left(\mathbf{r}', t - \frac{|\mathbf{r} - \mathbf{r}'|}{c}\right)\right\} d^3r' = 0, \quad (5.16)$$

where $R(|\mathbf{r}|)$ is the homogeneous, isotropic time-independent correlation of μ given by

$$\mathbf{E}\{\mu(\mathbf{r}_1, t_1)\mu(\mathbf{r}_2, t_2)\} = R(|\mathbf{r}_1 - \mathbf{r}_2|). \quad (5.17)$$

If $\mathbf{E}\{\psi(\mathbf{r}, t)\}$ is time harmonic with frequency ω,

$$\mathbf{E}\{\psi(\mathbf{r}, t)\} = e^{-i\omega t}\phi(\mathbf{r}), \qquad k_0 = \frac{\omega}{c},$$

$\phi(r)$ satisfies the equation

$$[\Delta + k_0^2(1 + \varepsilon^2\mathbf{E}\{\mu^2\})]\phi(\mathbf{r}) +$$

$$\frac{\varepsilon^2 k_0^2}{\pi c^2}\int\frac{\exp(ik_0|\mathbf{r} - \mathbf{r}'|)}{|\mathbf{r} - \mathbf{r}'|}\omega^2 R(|\mathbf{r} - \mathbf{r}'|)\phi(\mathbf{r}') \, d^3r' = 0. \quad (5.18)$$

For a time-independent medium this can be derived starting from the reduced wave equation itself (see Karal and Keller [32]). This equation points out that the random medium for small values of ε possesses a wave vector k given by

$$\left(\frac{k}{k_0}\right)^2 = 1 + \varepsilon^2\mathbf{E}\{\mu^2\} - 2i\varepsilon^2 k_0 \int_0^\infty (e^{2ik_0 r} - 1)R(r) \, dr. \quad (5.19)$$

An attenuation is introduced by the random medium, as evidenced by the imaginary part in (5.19). Since the real part of k is greater than the real part of k_0, the phase velocity is less in a random medium. This is understandable because random scattering increases the effective path in the medium.

A finite-order perturbation expansion of the average field in a random medium might lead to secular term difficulties. However, an infinite subseries of the original might give a nonsecular contribution. This has given rise to the method of summing up an infinite number of a class of diagrams in well-known situations in quantum electrodynamics, manybody theories, and statistical mechanics. This was adopted by Bourret [33] and a few others in the case of wave propagation in a random medium with the refractive index governed by a Gaussian type of probabilistic structure.

Multiple scattering of waves by randomly distributed scatterers in a continuous medium also give rise to situations which can be described by stochastic wave propagation. To take into account the individual scattering processes, suitable modification of the perturbation procedures used above has been proposed by Foldy [34], Lax [35], Twersky [36] who arrive at an effective field approximation which yields a suitable modification of the free space wave vector.

5.2. Schrodinger Equation with Stochastic Potentials

In quantum mechanics the solution of the Schrodinger equation with potentials describes the effects produced by the scattering of the particle from the potential. It is usual to decompose the wave function into a superposition of partial waves and write a separate Schrodinger equation for each of them. The effect of the potential is felt in terms of an additional phase term that appears in the asymptotic limit of the solution. These phase shifts are directly related to cross sections or probabilities associated with the scattering produced by the potential. If we deal with a potential which has a stochastic parameter in it, we can only talk in terms of the statistical averages of the solution. Hence we can talk of the mean and moments of the phase shifts.

As an example, we take an isotropic potential which is symmetric and which depends on the parameter r, the distance from the scattering centre. In this case the second-order one-dimensional Schrodinger equation can be written. Thus, the partial wave satisfies the equation

$$\psi''(l, k, r) + \left[k^2 - V(r) - \frac{l(l + 1)}{r^2} \right]\psi(l, k, r) = 0, \qquad (5.20)$$

where $V(r)$ is the random potential and k is the wave number. We also assume that the usual conditions imposed on the potential are satisfied almost

certainly in the sense of stochastic convergence. Then we can arrive at the
following differential equation for the phase function (see Calogero [37]):

$$\delta'(l, k, r) = - k^{-1}V(r)[\cos \delta(l, k, r)J_1(kr) - \sin \delta(l, k, r) N_1(kr)]. \qquad (5.21a)$$

We can certainly obtain the Born approximate solutions in a formal way as

$$\delta(l, k, r) = - k^{-1} \int_0^r V(r)J_l^2(kr)\, dr, \quad \text{or}$$

$$\delta(l, k) = - k^{-1} \int_0^\infty V(r)J_l^2(kr)\, dr. \qquad (5.21b)$$

We can also write an improved Born approximate solution in the form

$$\delta(l, k) = - k^{-1} \int_0^\infty dr V(r)J_l^2(kr) \exp\left[\frac{1}{2k} \int_r^\infty V(r')J_l(kr')N_l(kr')\, dr'\right]. \quad (5.22)$$

Finally let us impose a simple restriction on $V(r)$:

$$V = \frac{\mu}{r}, \qquad (5.23)$$

where μ is a random variable taking values ± 1, the probability per unit
distance of its transition from $+1$ to -1 and from -1 to $+1$ being p
and q, respectively. We can set up differential equations for $\pi(1, r)$ and
$\pi(-1, r)$:

$$\frac{\partial \pi(1, r)}{\partial r} = - p\pi(1, r) + q[1 - \pi(1, r)]$$

$$= - (p + q)\pi(1, r) + q \qquad (5.24)$$

with the conditions

$$\pi(1, 0) = 1, \qquad \pi(-1, 0) = 0,$$
$$\pi(1, r) + \pi(-1, r) = 1. \qquad (5.25)$$

The mean value of μ turns out to be

$$\bar{\mu} = \frac{q - p}{q + p} + \frac{2p}{p + q}e^{-(p+q)r}$$

so that

$$\bar{V}(r) = \frac{q - p}{q + p}\frac{1}{r} + \frac{2p}{p + q}\frac{e^{-(p+q)r}}{r}.$$

It is a pleasant surprise to note that the well-known Yukawa potential has a
stochastic interpretation. Starting with a Coulomb potential, we obtain the
average potential, which is partially Coulomb and partially Yukawa. Other
models of potentials have been investigated by Ramakrishnan, Vasudevan,

and Srinivasan [38]. Statistical properties of $\delta(l,\ k)$ in different approximations can be obtained from (5.22) and (5.24).

We wish to close this discussion with the note that the methods of solving the random wave propagation are far from complete. The methods employed are not quite rigorous, and the validity of the so-called "solutions" are to be tested in each case in terms of the underlying physical nature of the problem. However, for the physicist and the engineer, it is gratifying to note that many complex phenomena may have a simple statistical description which can be corroborated by experimental evidence.

REFERENCES

1. V. A. Ambartzumian, *Theoretical Astrophysics*, Pergamon Press, 1958, New York.
2. S. Chandrasekhar, *Radiative Transfer*, Oxford, Clarendon Press, 1950.
3. I. W. Bursbridge, *Mathematics of Radiative Transfer*, Cambridge University Press, 1960.
4. R. Bellman and R. Kalaba, *Proc. Nat. Acad. Sci., U.S.*, **42**(1956), 629; **43**(1957), 517.
5. T. E. Harris, *Theory of Branching Processes*, Springer-Verlag, Berlin, 1963.
6. Alladi Ramakrishnan, Probability and Stochastic Processes, in *Handbuch der Physik*, Vol. 3, Springer-Verlag, Berlin, 1959.
7. R. Bellman and T. E. Harris, *Proc. Nat. Acad. Sci. U.S.*, **34**(1948), 601.
8. G. Milton Wing, *An Introduction to Transport Theory*, John Wiley, New York, 1962.
9. J. B. Keller, in *Proc. Symp. Appl. Math.*, Vol. 13, Amer. Math. Soc., Providence, R.I., 1962.
10. U. Frisch, *Probability Methods in Applied Mathematics*, Vol. 1, Academic Press, New York, 1968.
11. R. H. Kraichnan, *J. Math. Phys.*, **2**(1961), 124.
12. V. I. Tatarski, *Wave Propagation in Turbulent Medium*, McGraw-Hill, New York, 1961.
13. B. Davison, *Neutron Transport Theory* Oxford, Clarendon Press, 1957.
14. A. M. Weinberg and E. P. Wigner, *The Physical Theory of Neutron Chain Reactors*, University of Chicago Press, Chicago, Ill., 1958.
15. L. Takacs, *Publ. Math. Inst. Hung. Acad. Sci.*, **1**(1956), 55.
16. E. Hopf, *Mathematical Problems of Radiative Equilibrium*, Cambridge, Univ. Press Camb. Tracts in Math. and Physics, No. 31, 1934.
17. K. M. Case, *Ann. Phys.*, **9**(1960), 1.
18. E. P. Wigner, *Proc. Symp. Appl. Math.*, **13**(1959), 89, Amer. Math. Soc., Providence, R.I.
19. J. H. Tait, *Rept. Progr. Physics*, **19**(1956), 268.
20. Alladi Ramakrishnan, R. Vasudevan, and P. Rajagopal, *J. Math. Anal. and Appl.*, **1**(1960), 145.
21. Alladi Ramakrishnan and S. K. Srinivasan, *Bull. Math. Biophys.*, **20**(1958), 288.
22. R. Bellman, R. Kalaba, and G. M. Wing, *Proc. Nat. Acad. Sci. U.S.*, **43**(1957), 517; **7**(1958), 149, 741, **8**(1959), 249, 545.
23. Alladi Ramakrishnan, *Proc. Cambridge Philos. Soc.*, **46**(1950), 595.
24. R. Bellman, R. Kalaba, and R. Vasudevan, *J. Math. Anal. Appl.*, **8**(1964), 225.
25. R. Bellman, R. Kalaba, and R. Vasudevan, *J. Math. Anal. Appl.*, **7**(1963), 264.

26. S. Ueno, *Astrophys. J.*, **126**(1957), 413.
27. R. E. Bellman and R. Kalaba, *Proc. Nat. Acad. Sci. U.S.*, **42**(1956), 629.
28. R. E. Bellman, R. E. Kalaba, and M. C. Prestrud, *Invariant Imbedding and Radiative Transfer in Slabs of Finite Thickness*, American Elsevier, New York, 1963.
29. J. B. Keller, *Proc. Symp. Appl. Math.*, Vol. 16, Amer. Math. Soc., Providence, R.I., 1964, p. 145.
30. R. H. Kraichnan, *Proc. Symp. Appl. Math.*, Vol. 13, Amer. Math. Soc., Providence, R.I. (1962), p. 199.
31. J. Richardson, *Proc. Symp. Appl. Math.*, Vol. 16, Amer. Math. Soc., Providence, R.I., 1964, p. 290.
32. F. C. Karal and J. B. Keller, *J. Math. Phys.*, **5**(1964), 537.
33. R. C. Bourret, *Nuovo Cimento*, **26**(1962), 1.
34. L. L. Foldy, *Phys. Rev.* (2) **67**(1945), 107.
35. M. Lax, *Rev. Mod. Phys.*, **23**(1951), 287.
36. V. Twersky, *Proc. Symp. Appl. Math.*, Vol. 16, Amer. Math. Soc., Providence, R.I., p. 84.
37. F. Calogero, *Nuovo Cimento*, **27**(1963), 261.
38. Alladi Ramakrishnan, R. Vasudevan, and S. K. Srinivasan, *Z. für Phys.*, **196**(1966), 112.

Chapter 8

PATH INTEGRALS IN
CLASSICAL AND QUANTUM PHYSICS

1. INTRODUCTION

More than forty years have passed since the Schrodinger equation, which governs the evolution in time of wave functions describing microscopic particles, was formulated. Even though the physical content of these wave functions rests heavily on Born (probabilistic) interpretation [1, 2] of Quantum Mechanics, a direct link between stochastic processes which deal with the time development of probabilities and a quantum mechanical description of nature is yet to emerge. Although formal similarities of certain aspects of both have been noticed and elucidated, as in the case of the diffusion and the Schrodinger equations, the solution ψ of the Schrodinger equation does not lend itself to a probabilistic interpretation directly. To develop a probabilistic description of quantum mechanics, it is necessary to introduce the probability density $\rho = \psi^*\psi$. One striking difference between stochastic theories and quantum mechanics arises from the distinct type of super-position principle that is employed. In stochastic theory, probabilities are superposed, while in quantum mechanics the amplitudes or wave functions are linearly superposed. To establish a contact between quantum and probability theories, various studies involving the use of different types of phase space distributions (synthesized with the use of wave functions) like the Wigner distribution [3] have been carried out. These attempts are effectively equivalent to incorporating a certain amount of coarse graining into the quantum mechanical description of nature.

It was Feynman [4], using the Chapman–Kolmogorov method, who pointed out the clear-cut analog that is found to exist in the derivations of the Schrodinger equation and the equation of motion for the Brownian particle. However, as we have observed in the study of Brownian motion, the analog is only the evolution of the complex amplitudes that behave like probabilities. Hence the analogy between these two theories is only formal.

An analysis of the exact relation between quantum mechanical laws and classical stochastic theory has been presented by Ramakrishnan, Vasudevan, and Ranganathan [5, 6]. The Feynman paths and their probability content as well as the underlying physical concepts have been brought forth in detail. There have been other attempts [7] to illustrate that the uncertainty principle has a stochastic analog. Using only classical mechanical concepts, it was shown that an ensemble of charged harmonic oscillators may be viewed as indistinguishable from its quantum mechanical counterpart. The only extra hypothesis needed is a classical random electromagnetic field with specific properties. In this chapter we present a brief account of Feynman's derivation [8] of the Schrodinger equation from path integral formalism and discuss a few important applications of the method.

In Section 2 we introduce the path probability. Then we discuss Feynman's derivation of the Schrodinger equation. The classical harmonic oscillator is discussed from the viewpoint of path integrals, and we demonstrate how such a classical oscillator can simulate its quantum mechanical counterpart. Section 3 deals with the problems of Statistical Physics, a potential area from the current viewpoint.

2. PATH INTEGRAL FORMALISM

As we mentioned in Section 1, path integral formalism was developed by Feynman following some remarks of Dirac [9] concerning the relation of classical action to quantum mechanics. Since the formulation to be presented contains as its essential idea the concept of probability amplitude associated with a completely specified motion as a function of time, the principle of superposition of probability amplitudes will play a central role. In Section 2.1 we present a brief account of superposition principle. We then proceed to the formulation of path integral.

2.1. Superposition Principle

The principle of superposition of probabilities in quantum mechanics can be stated as follows:

If events A, B, and C occur and if P_{ab} is the probability that, if A gives the result a, B gives the result b, we have

$$P_{ac} = \sum_b P_{ab}P_{bc}, \tag{2.1}$$

provided the events B and C are causally independent. In quantum mechanics there exist complex amplitudes ψ_{ab} such that $P_{ab} = |\psi_{ab}|^2$ and $P_{bc} = |\psi_{bc}|^2$. However, if no attempts are made to measure B in quantum mechanics, the equation (2.1) of classical probability theory is to be replaced by

$$\psi_{ac} = \sum_b \psi_{ab}\psi_{bc}. \tag{2.2}$$

Once an experiment is performed, the system gets disturbed and we are dealing with a new system, so that the law of superposition is different. In fact, if we know, by a measuring device, that the system is in state B, the probability P_{ac} is uniquely determined by the classical law. If the ψ_{ab}'s are taken to be wave amplitudes associated with particles like electrons or photons, the intensity of these waves at a point measures the probability of finding the particles at these locations.

Equation (2.2) points out that any determination of the choices taken by a process capable of following different routes destroys the interference between the amplitudes of these choices. This is the essence of Heisenberg's uncertainty principle [10]. In other words, the operation of measuring equipment, however subtle it may be, is sufficient to disturb the system and its probability amplitude. Thus we have the result that there is a natural limit to the subtlety of any experiment or the refinement of any measurement. We finally note that (2.2) can be easily extended to a sequence of events $A, B, C, D, E, F, \ldots$.

2.2. Path Integral

If a particle can assume values for its position coordinates $x_1, x_2, \ldots, x_i, \ldots, x_n$ at times denoted by $t_1, t_2, \ldots, t_i, \ldots, t_n$ so that $t_{i+1} = t_i + \varepsilon$, where ε is a small interval, we can think of a path $x(t)$ of the particle. We can then associate with it a complex function $\phi(x_1, \ldots, x_i, \ldots, x_n)$ denoting the probability amplitude for this specified path. The probability amplitude for the particle to lie in a given region R is $\psi(\text{R})$ given by

$$\psi(\text{R}) = \lim_{\varepsilon \to 0} \int_{\text{R}} \phi(x_1, x_2 \ldots, x_n)\, dx_1\, dx_2, \ldots, dx_n, \tag{2.3}$$

the probability itself being given by the modulus square of the amplitude $\psi(\text{R})$.

Having introduced the notion of a path probability, for a particle to go from a point a at time t_a to a point b at time t_b a quantum mechanical rule for the contributions arising from each trajectory in going from a to b can be specified.

The probability amplitude $K(b, a)$ in this case is given by

$$K(b, a) = \sum_{\text{all paths}} \phi[x(t)]. \tag{2.4}$$

To evaluate $\phi[x(t)]$ Feynman postulates that the paths contribute equally in magnitude, but the phase of their contribution is the classical action in units of $\hbar$, that is, the time integral of the Lagrangian taken along the path. Thus ϕ, which is now a functional of the path $x(t)$, is given by

$$\phi[x(t)] = \text{constant } e^{iS[x(t)]/\hbar}, \tag{2.5}$$

S being the classical action along that path which, as usual, is the extremum of the time integral over the Lagrangian:

$$S[x(t)] = \int_{t_a}^{t_b} L(\dot{x}, x, t)\, dt. \tag{2.6}$$

The normalizing constant in (2.5) is to be chosen conveniently to enable us to go over to the conventional formulation of quantum mechanics. To understand the physical meaning of (2.5), we should remember that $\hbar$ is a very small quantity ($\hbar = 1\cdot05 \times 10^{-27}$ erg sec) and S is a large quantity, so that $e^{iS/\hbar}$ will oscillate violently, the contributions from neighboring paths getting canceled. However, for the special path, which is the classical trajectory, the action is an extremum and small changes in the path produce in S no change at least in the first order. All paths in the neighborhood of the classical path do contribute in the same phase and constructively interfere. Thus trajectories in the vicinity of the classical path are important as long as the action is still within about $\hbar$ of the S classical. Also, in the classical limit when $\hbar$ tends to zero and S is very large, the extremal classical trajectory is the only valid path of the particle for a given Lagrangian.

To obtain the sum of the contributions arising from different paths, as in Riemann integration, we choose a subset of all such paths. We divide the time interval into N steps of ε each and define positions x_i corresponding to each time instant t_i such that $t_0 = t_a$, $x_0 = x_a$, $t_N = t_b$, $x_N = x_b$, and $\varepsilon = t_{i+1} - t_i$, so that $N\varepsilon = t_b - t_a$. We can connect these points to represent one path and integrate over all the points x_i except over the end points x_a and x_b. This process can be refined further and further by defining paths corresponding to smaller values of ε. As in the case of the Riemann integral, the normalizing factors anticipated in (2.5) should depend on ε. For Lagrangians depending only on x and $\dot{x}$ up to the second power, this factor turns out to be A^{-N}, where $A = (2\pi i\hbar\varepsilon/m)^{\frac{1}{2}}$. Hence the amplitude $K(b, a)$ is given by

$$K(b, a) = \lim_{\varepsilon \to 0} \frac{1}{A} \iint \cdots \int \exp \frac{i}{\hbar} \sum_{i=1}^{N-1} S(x_{i+1}, x_i) A^{-(N-1)}\, dx_1\, dx_2 \ldots dx_{N-1},$$

$$\tag{2.7}$$

where

$$S(x_{i+1}, x_i) = \min \int_{t_i}^{t_{i+1}} L(\dot{x}(t), x(t))\, dt. \tag{2.8}$$

This concept of sum over all paths should be compared with the definitions and rules given for the integration procedures developed in Section 2.1 of Chapter 2. In a more general notation, the sum over all paths can be written as

$$K(b, a) = \int_a^b e^{(i/\hbar)S[b,a]}\, \mathscr{D}(x(t)), \tag{2.9}$$

and this is called the path integral. This integral over trajectories was introduced by Wiener [11] when he studied Brownian motion.*

As a simple example, let us consider a free particle moving in one dimension defined by the Lagrangian $L = m\dot{x}^2/2$. The kernel function $K(b, a)$ is given by

$$K(b, a) = \lim_{\varepsilon \to 0} \left(\frac{2\pi i\hbar t}{m}\right)^{-N/2} \int \cdots \int \exp \frac{im}{2\hbar\varepsilon} \sum_{i=1}^{N} (x_i - x_{i-1})^2 \, dx_1 \, dx_2 \ldots dx_{n-1}$$

$$= \left[\frac{2\pi i\hbar(t_b - t_a)}{m}\right]^{-\frac{1}{2}} \exp \frac{im(x_b - x_a)^2}{2\hbar(t_b - t_a)}. \tag{2.10}$$

The composition of the kernels for successive events in time is quite a simple matter, since the action along the path in the phase adds up at each time. It is not difficult to check that, if we split up the path between b and a at an intermediate time t_c and position x_c, we can multiply the amplitudes for events occurring in succession in time to obtain

$$K(b, a) = \int K(b, c)K(c, a) \, dx_c. \tag{2.11}$$

Though the functional integrations and the path integral formulation of Feynman were similar to the Wiener integrals, pure mathematicians like Ito [16], Cameron [17], and Gilson [18] undertook the task of giving rigorous proofs to Feynman's results. Difficulties occur because of the nature of Feynman measure. Gilson [19] presented a rigorous approach and discussed special properties of the kernel. He also points out a method of obtaining Wigner distribution from Feynman's propagation kernel. However, it is to be noted that Wigner distribution (see, for example, Mori *et al.* [20]) is a sort of coarse-grained average performed on a quantum mechanical system, and it gives the joint probability distribution in both momentum and coordinate space, which is not at all a possible description in an exactly quantum mechanical picture because of the uncertainty principle. The possibility of obtaining quantum behavior of systems which can be deterministic or classical in description has been investigated by Moyal [21] and others who have come to the conclusion that such quantum systems can exist. The harmonic oscillator provides a simple example of such a quantum mechanical system.

3. APPLICATIONS OF PATH INTEGRALS

In this section we demonstrate the usefulness of path integral formalism. The best example is the derivation of the Schrodinger equation and its

---

* The mathematical aspects of the connection between the path integrals employed in Brownian motion and quantum mechanics have been the subject of study by Kac [12]. Montroll [13], Gelfand and Yaglom [14], and Saito and Namiki [15].

properties, as spectacularly demonstrated by Feynman [4]. We discuss this in Section 3.1. In Section 3.2 we derive some of the interesting properties of a classical harmonic oscillator. Section 3.5 is devoted to general problems in Statistical Physics and various viewpoints on the general philosophy of the formalism.

3.1. Schrödinger Equation

Our starting point is the kernel function defined in Section 2.2. It is quite easy to go over to the definition of a quantum mechanical wave function from the concept of the kernel K, defining the amplitude for the particle as going from the space time point (x_1, t_1) to the space time point (x_2, t_2). Thus the wave function $\psi(x, t)$ for the particle at the space time point (x_1, t_1) is the amplitude for arriving at that particular space time point without any knowledge of the past motion. The modulus square of the wave function $|\psi(x, t)|^2$ yields the probability of finding the particle at the space time point (x, t). Since the wave function is only an amplitude in the same sense as K is, we can use the concept of composition of amplitudes. The wave function ψ satisfies

$$\psi(x_2, t_2) = \int_{-\infty}^{+\infty} K(x_2, t_2; x_1, t_1)\psi(x_1, t_1)\, dx_1, \tag{3.1}$$

an equation similar to Smoluchowsky's equation for Markov processes. Equation (3.1) implies that all past history is irrelevant for arriving at the amplitude necessary for the particle to be at the space time point (x_2, t_2) provided we are given $\psi(x_1, t_1)$.

From (3.1) it is easy to derive the Schrödinger differential equation by considering the development of the wave function in the time interval $(t, t + \varepsilon)$. For purposes of illustration let us assume that the motion is one-dimensional and that the particle is subjected to a potential $V(x, t)$. The Lagrangian governing the motion is given by

$$L = \frac{m\dot{x}^2}{2} - V(x, t). \tag{3.2}$$

The action for the short interval ε can be taken to be ε times the Lagrangian and velocity $\dot{x} = \Delta x/\varepsilon$. By virtue of (3.1), we have

$$\psi(x, t + \varepsilon) =$$
$$\int_{-\infty}^{\infty} \frac{1}{A} \exp\left[\frac{i}{\hbar}\frac{m(x - y)^2}{\varepsilon}\right] \exp\left[\left(\frac{-i}{\hbar}\right)\varepsilon V\left(\frac{x + y}{2}, t\right)\right]\psi(y, t)\, dy. \tag{3.3}$$

11*

Since the integral over y gives contributions only if x is near y, we substitute $y = x + \delta$ and expand ψ in a power series around x to obtain

$$\psi(x, t) + \varepsilon\frac{\partial\psi(x, t)}{\partial t} =$$

$$\int_{-\infty}^{\infty} \frac{1}{A} e^{im\delta^2/(2\hbar\varepsilon)} \left[\psi(x, t) + \delta\frac{\partial\psi(x, t)}{\partial x} + \frac{\delta^2}{2}\frac{\partial^2\psi(x, t)}{\partial x^2} \right] e^{-i\varepsilon V(x)/\hbar}\, d\delta. \quad (3.4)$$

After some manipulation, we are led to the usual Schrodinger equation for a particle moving in one direction under the action of a potential $V(x, t)$:

$$-\frac{\hbar}{i}\frac{\partial\psi(x, t)}{\partial t} = \left[\frac{1}{2m}\left(\frac{\hbar}{i}\frac{\partial}{\partial x}\right)^2 + V(x, t)\right]\psi(x, t) = H\psi(x, t), \quad (3.5)$$

where H is the Hamiltonian of the system. It is easy to show that the kernel function $K(b, a)$ obeys the same Schrodinger equation for all times $t_b > t_a$ and is zero for $t_b < t_a$:

$$-\frac{\hbar}{i}\frac{\partial K(b, a)}{\partial t_b} = -\frac{\hbar^2}{2m}\frac{\partial^2}{\partial x_b^2}K(b, a) + V(x_b, t_b)K(b, a), \quad t_b > t_a,$$

$$K(b, a) = 0 \quad \text{for} \quad t_b < t_a. \quad (3.6)$$

In a more general context, $K(b, a)$ satisfies the equation

$$-\frac{\hbar}{i}\frac{\partial K(b, a)}{\partial t_b} - H_b K(b, a) = -\frac{\hbar}{i}\delta(x_b - x_a)\delta(t_b - t_a). \quad (3.7)$$

The stationary solution ψ_n of the Schrodinger equation (3.5) can be written

$$\psi_n = e^{-iE_n t/\hbar}\,\phi_n(x), \quad (3.8)$$

where $\phi_n(x)$ satisfies the equation

$$H\phi_n = E_n\phi_n \quad (3.9)$$

and E_n is the eigenvalue of energy corresponding to the nth stationary state or the eigensolution ϕ_n of the Schrodinger equation. Using the orthonormal properties of the eigensolutions ϕ_n, the propagation kernel K can be written

$$K(x_2, t_2, x_1, t_1) = \sum_{n=1}^{\infty} \phi_n(x_2)\phi_n^*(x_1)e^{(-i/\hbar)E_n(t_2 - t_1)} \quad \text{for} \quad t_2 > t_1,$$

$$= 0 \quad \text{for} \quad t_2 < t_1. \quad (3.10)$$

3.2. Harmonic Oscillator

One of the interesting problems in quantum mechanics that can be completely analyzed by this formalism is the quantum mechanical harmonic oscillator defined by the Lagrangian

$$L = \frac{m}{2}(\dot{x}^2 - \omega^2 x^2). \quad (3.11)$$

For a time difference $t_b - t_a = T$, the classical action along the path is given by

$$S_{cl} = \frac{m\omega}{2 \sin \omega T}[(x_a{}^2 + x_b{}^2) \cos \omega T - 2x_a x_b]. \tag{3.12}$$

The evaluation of path integrals is usually difficult. However, in this case it is advantageous to expand the path $x(t)$ around the classical path $\bar{x}(t)$. This device, due to Davison [22] and Burton and Borde [23], renders the evaluation fairly easy. Thus in the path integral given by

$$K(b, a) = \int_a^b \exp\left\{\frac{i}{\hbar} \int_{t_a}^{t_b} \frac{m}{2}(\dot{x}^2 - \omega^2 x^2)\, dt\right\} \mathscr{D}(x(t)) \tag{3.13}$$

we replace x by $\bar{x} + y$, where $\bar{x}$ is the classical path and y represents the deviations from it. Since only small deviations really contribute, it is enough to expand the functions up to y^2 order. Thus $S = S_{cl} +$ terms second-order in y. The term $e^{S_{cl}(i/\hbar)}$ separates out and contains $V(y)$ and $V(\bar{x})$ only. The remaining integral contains y, which goes from 0 to 0 and in time 0 to T. The result can therefore be separated into two factors, one of them being purely a function of time T:

$$K(b, a, T) = F(T) \exp\left\{\frac{im\,\omega}{2\hbar \sin \omega T}[(x_a{}^2 + x_b{}^2) \cos \omega T - 2x_a x_b]\right\}, \tag{3.14}$$

where

$$F(T) = \int_0^0 \exp\left\{\frac{i}{\hbar} \int_0^T \frac{m}{2}(\dot{y}^2 - \omega^2 y^2)\, dt\right\} \mathscr{D}(y(t)). \tag{3.15}$$

A method for handling the integral in (3.15) is to expand y in Fourier series, since it goes from $y = 0$ at $t = 0$, to $y = 0$ at $t = T$:

$$y = \sum_n a_n \sin \frac{n\pi t}{T}. \tag{3.16}$$

After some calculations we arrive at the result

$$F(T) = \left(\frac{m\omega}{2\pi i\hbar \sin \omega T}\right)^{\frac{1}{2}}. \tag{3.17}$$

A knowledge of K provides an alternative method of obtaining the eigenvalues and wave functions of the harmonic oscillator without actually solving the Schrodinger equation, because the kernel function itself is expressible in terms of all the solutions ϕ_n and $\phi_n{}^*$ for the stationary case. $K(b, a)$, given in (3.13), can be rewritten

$$K(x_1, x_2; T) = \left(\frac{m\omega}{\pi\hbar}\right)^{\frac{1}{2}} e^{-i\omega T/2} (1 - e^{-2i\omega T})^{\frac{1}{2}} \cdot$$

$$\exp\left\{ -\frac{m\omega}{2\hbar}\left[(x_1{}^2 + x_2{}^2)\left(\frac{1 + e^{2i\omega T}}{1 - e^{2i\omega T}}\right) - \frac{4x_1 x_2 e^{i\omega T}}{(1 - e^{2i\omega T})}\right]\right\}$$

$$= \sum_{n=0}^{\infty} e^{(-i/\hbar)E_n T} \phi_n(x_2)\phi_n{}^*(x_1). \tag{3.18}$$

Looking at the left-hand side of (3.18), we notice that, because of the factor $e^{-i\omega T/2}$, all terms in the expansion will be of the form $e^{-i(n+\frac{1}{2})\omega T}$ for $n = 0$, $1, 2, 3, \ldots$. This indicates that the energy levels are given by

$$E_n = (n + \tfrac{1}{2})\hbar\omega. \tag{3.19}$$

We can also show that the lowest eigenwave function corresponding to energy $E_0 = \frac{1}{2}\hbar\omega$ of the ground state of the oscillator is given by

$$\phi_0(x) = \left(\frac{m\omega}{\pi\hbar}\right)^{\frac{1}{4}} e^{-m\omega x^2/2\hbar}. \tag{3.20}$$

Proceeding to the next higher order, we arrive at the first excited state corresponding to the energy $E_1 = \frac{3}{2}\hbar\omega$, and the eigenwave function $\phi_1(x) = (2m\omega/\hbar)x\phi_0(x)$. The second excited state obtained by equating the corresponding coefficients on both sides is characterized by the energy value $E_2 = \frac{5}{2}\hbar\omega$, and the wave function $\phi_2(x) = (1/\sqrt{2})[(2m\omega/\sqrt{2})x^2 - 1]\phi_0(x)$. This procedure offers a method of solving for the eigenstates of a quantum mechanical system.

3.3. Transition Amplitude

To get a deeper understanding of quantum mechanical processes it is necessary to introduce the concept of transition probability in quantum mechanics. Let us picture a system in the state described by the wave function $\psi(x_1, t_1)$ evolving with respect to t, and let us measure it by an experiment characterized by the state represented by $\chi(x_2, t_2)$. Here we are interested in the probability of finding the system in starting from the original state $\psi(x_1, t_1)$:

$$P_{\chi\psi} = |\int \chi^*(x_2, t_2)\phi(x_2, t_2)\, dx_2|^2, \tag{3.21}$$

where ϕ is the state to which $\psi(x_1, t_1)$ has developed in the time interval $(t_2 - t_1)$. This can be obtained by the application of the propagation kernel to carry the system from t_1 to t_2 under the influence of the given forces acting on the system. The transition amplitude can be explicitly written as (see Feynman [4])

$$\langle\chi|\psi\rangle = \int \chi^*(x_2, t_2)K(2, 1)\psi(x_1, t_1)\, dx_2\, dx_1. \tag{3.22}$$

By inserting the expressions relevant to the problem under study, $K(2, 1)$ can be expressed in terms of the action along the paths traversed, and hence we have

$$\langle \chi | \psi \rangle = \iint \chi^*(x_2, t_2) e^{iS/\hbar} \, \psi(x_1, t_1) \mathscr{D}(x(t)) \, dx_1 \, dx_2. \tag{3.23}$$

Equation (3.23) is the starting point of all calculations leading to computation of cross section for quantum mechanical processes, and in many cases the kernel $K(1, 2)$ can be expanded in a power series in the interaction and gives rise to the various Born terms in the perturbation series. In each order of perturbation these terms are usually represented by diagrams called Feynman diagrams. The perturbation techniques have led to very important results in the field of Quantum Electrodynamics; they are described exhaustively in various textbooks (see, for example, Ramakrishnan [24]).

3.4. Brownian Motion

As indicated in Chapter 4, methods of measure theory and integration in functional spaces are widely applied in theories of random processes, and they are related to path integrals corresponding to trajectories of Brownian particles. This measure in the space of continuous functions was introduced by Wiener [11] in the 'twenties. Kac [12] considered in great detail ways of calculating the mean value of certain functionals over Brownian paths. If we examine the path of a particle undergoing Brownian motion along the x-axis under random impulses starting from the region at time $t = 0$, ignoring the inertia of the particle, the probabilities for its position at times $t \neq 0$ will be given, as explained in Chapter 4, by the diffusion equation

$$\frac{\partial \pi(x, t)}{\partial t} = D \frac{\partial^2 \pi(x, t)}{\partial x^2}, \tag{3.24}$$

where D is the diffusion coefficient given by the well-known Einstein relations. The probability density is given by

$$\pi(x, t) = \frac{1}{(4\pi D t)^{\frac{1}{2}}} \exp\left(-\frac{x^2}{4Dt} \right). \tag{3.25}$$

Assuming, for simplicity, that $D = \frac{1}{4}$, the probability that the coordinate $x(t_1)$ of the particle is found between a_1 and b_1 at time t_1, between a_2 and b_2 at time $t_2, \ldots$, and between a_n and b_n at time t_n can be written as

$$[\pi^n t_1(t_2 - t_1) \ldots (t_n - t_{n-1})]^{-\frac{1}{2}} \cdot$$

$$\int_{a_1}^{b_1} \int_{a_2}^{b_2} \cdots \int_{a_n}^{b_n} \exp\left[-\frac{x_1^2}{t_1} - \frac{(x_2 - x_1)^2}{t_2 - t_1} \cdots - \frac{(x_n - x_{n-1})^2}{t_n - t_{n-1}} \right] dx_1 \ldots dx_n. \tag{3.26}$$

Wiener considered the class $\mathscr{C}$ of real functions $x(t)$ continuous in the interval $0 < t < 1$. He also introduced the quasi-interval as the set of all of the functions in $\mathscr{C}$ satisfying the relations $a_i < x(t_i) < b_i$, $i = 1, \ldots, n$, where a_i, b_i are bounded real numbers. The Wiener measure of the quasi-interval is defined as the expression (3.26), the measure having all the abstract properties of the Lebesgue measure. The Wiener integral of a functional $F[x(t)]$ of $x(t)$ over the entire class $\mathscr{C}$ is defined by $\int F[x, (t)] \, d_w x$. If we consider the interval 0 to t, rather than 0 to 1, and assume for simplicity that this interval is split into n equal parts, the Wiener integral is written as

$$\int F(x) \, d_\omega x = \lim_{n \to \infty} \left(\frac{n}{\pi t}\right)^{n/2} \int_{-\infty}^{\infty} \cdots \int_{-\infty}^{\infty} F(x_1 \ldots x_n) \cdot$$

$$\exp - \left(\frac{n}{t}\right)[x_1 + (x_2 - x_1)^2 + \ldots + (x_n - x_{n-1})^2] \, dx_1 \ldots dx_n, \quad (3.27)$$

where $x_m = it/m$. For more mathematical details and proofs for the existence of such integrals and their properties, the reader is referred to the survey article by Gelfand and Yaglom [14]. The simplest case that can be considered is one in which the functional $F[x(t)]$ is the product of the values of the function $x(t)$ at a finite number of points. By definition the integral is nothing but a multidimensional integral. The integrations can now be performed easily, and we obtain the well-known moments of the Brownian process:

$$\int x(t) \, d_w x = 0,$$

$$\int x(t_1)x(t_2) \, d_w x = b(t_1, t_2),$$

$$\int x(t_1)x(t_2) \ldots x(t_{2n+1}) \, d_w x = 0,$$

$$\int x(t_1)x(t_2) \ldots x(t_{2n}) \, d_w x = \sum_{\text{all pairs}} b(t_{i_1}, t_{i_2})b(t_{i_3}, t_{i_4}) \ldots \quad (3.28)$$

More interesting is the expectation value of the functional $F[x(t)]$ given by

$$F[x(t)] = \exp \lambda \int_0^t x^2(\tau) \, d\tau. \quad (3.29)$$

The Wiener integral in this case has been calculated by ingenious methods [2] and is given by

$$\mathbf{E}[F(x)t))] = \int \exp \lambda \int x^2(\tau) \, d\tau \, d_w x$$

$$= [\sec t(\lambda)^{\frac{1}{2}}]^{\frac{1}{2}} \quad \text{for} \quad \lambda^{\frac{1}{2}} < \pi/2t. \quad (3.30)$$

Conditional Wiener measures and integrals have been considered which deal with situations in which $x(0) = 0$, and $x(t) = x$, and integrations are performed over all x_i's except the last x_n. Using this idea, evaluation of a functional of the form

$$F[x(t)] = \exp - \int_0^t V[x(\tau)] \, d\tau \tag{3.31}$$

over the Wiener measure can be performed to yield $\psi(x, t)$ given by

$$\psi(x, t) = \int \exp\left[- \int_0^t V[x(\tau)] \, d\tau \right] d_w x. \tag{3.32}$$

It can then be shown that $\psi(x, t)$ obeys a parabolic differential equation of the form

$$\frac{\partial \psi(x, t)}{\partial t} = \tfrac{1}{4} \frac{\partial^2 \psi(x, t)}{\partial x^2} - V(x)\psi(x, t). \tag{3.33}$$

This, as one might easily recognize, is similar to the Schrodinger equation arrived at by Feynman, of course without the appearance of the factors like i and $\hbar$, which are peculiar to quantum mechanical situations.

Before concluding this section, we wish to observe the possibility of representing the solutions of the differential equations of the type (3.33) in the form of a functional integral. Writing the differential equation as a difference equation and starting from the initial value, the value at a point (x, t) appears as a sum extending over all preceding points of the path. This procedure may be carried on till the last point, and a limiting process of the contributions from all possible continuous paths can be thought of. In this fashion the solution to a wide class of problems may possibly be achieved, and such a procedure may be of considerable interest from the viewpoint of the theory of differential equations.

3.5. Statistical Physics

In quantum statistical mechanics, we define a quantity called the partition function, Z, expressed in terms of the various extensive parameters. All the standard thermodynamic quantities like entropy, interval energy, and pressure are to be derived from a knowledge of Z and its differentials. To arrive at Z we define an entity called the density matrix and given by

$$\rho(x', x) = \sum_n \phi_n(x')\phi_n{}^*(x)e^{-\beta E_n}, \tag{3.34}$$

where $\beta = 1/kT$, T being the absolute temperature and k the Boltzman constant. The probability $\rho(x)$ of observing the system in x is given by the absolute square of the normalized amplitude, which is given by

$$\rho(x) = \frac{1}{Z} \sum_n \phi_n{}^*(x)\phi_n(x)e^{-\beta E_i}, \tag{3.35}$$

where Z is the partition function defined by

$$Z = \int \rho(x, x)\, dx \qquad (3.36)$$

$$= \text{Trace } \rho.$$

Let us recall the definition of propagation kernel given by

$$K(x_2, t_2; x_1, t_1) = \sum_i \phi_i(x_2)\phi_i{}^*(x_1) \exp[-(i/\hbar)E_i(t_2 - t_1)]. \qquad (3.37)$$

Comparing (3.36) and (3.37), we realize that, if the time difference $t_2 - t_1$ of (3.37) is replaced by $-i\beta\hbar$, the expression for the density matrix $\rho(x_2, x_1)$ is identical with the expression for the kernel. Writing the density matrix in an explicit fashion, we have

$$K(x_2, u_2; x_1, u_1) = \sum_i \phi_i(x_2)\phi_i{}^*(x_1) \exp\{-[u_2 - u_1]E_i/\hbar\}, \qquad (3.38)$$

where u_1 and u_2 are the initial and final temperatures. Differentiating this expression with respect to u_2, we obtain

$$-\hbar\frac{\partial K}{\partial u_2} = \sum_i E_i\phi_i(x_2)\phi_i{}^*(x_1) \exp\{-([u_2 - u_1]E_i/\hbar)\}. \qquad (3.39)$$

Since $H_2\phi_i(x_2) = E_i\phi_i(x_2)$, we obtain the equation of motion (see [25, 26]) of the density matrix,

$$-\frac{\partial \rho(2, 1)}{\partial \beta} = H_2\rho(2, 1) \quad \text{for} \quad u_2 > u_1. \qquad (3.40)$$

The composition law for the kernels for successive intervals of u is the same as before and, dividing $u_2 - u_1$ into N small intervals each of length δ for a system with a Hamiltonian $H = (\hbar^2/2m)(d^2/dx^2) + V(x)$, we obtain the path integral for the kernel as

$$K(x_2, u_2; x_1, u_1) = \int \left(\exp\left\{ -\sum_1^{N-1} \frac{m}{2\hbar\delta}(x_{i+1} - x_i)^2 + \frac{\delta}{\hbar}V(x) \right\} \right) \prod_{i=1}^N \frac{dx_i}{a}. \qquad (3.41)$$

The normalizing factor a in this case is given by $a = 2\pi\hbar\delta/m$, and $u_2 - u_1 = N\delta$. If the path $x(u)$ is defined in terms of the parameter u, and if we call $\dot{x} = dx/du$ the density matrix is given by the path integral

$$\rho(x_2, x_1) = \int \exp\left\{ -\frac{1}{\hbar} \int_0^{\beta\hbar} \left[\frac{m}{2}(\dot{x}(u))^2 + V(x) \right] du \right\} \mathscr{D}[x(u)]. \qquad (3.42)$$

In other words, we should consider all possible paths or motions of the system by which the system goes from the initial to the final state in time $\beta\hbar$. The density matrix is the sum of the contributions from all the paths, some being large and others small. There is no cancellation from different paths. Using this idea, the density matrix for a harmonic oscillator is easily obtained as

$$\rho(x', x) = \left(\frac{m\omega}{2\pi\hbar \, \sinh \, \omega\beta\hbar}\right)^{\frac{1}{2}} \cdot$$

$$\exp\left\{-\frac{m\omega}{2\hbar(\sinh \, \omega\beta\hbar)^2} \left[(x^2 + x'^2) \cosh \, \beta\hbar\omega - 2x'x\right]\right\}. \quad (3.43)$$

Evaluating the partition function $Z = \text{Trace } \rho$, we arrive at the well-known expression for the free energy of an oscillator:

$$F = kT \log \left[\sinh(\hbar\omega/2kT)\right]. \quad (3.44)$$

The construction of path integrals for evaluating partition functions have been effectively used by Feynman [27] in studying liquid helium and the polaron problem in crystals. Recent developments in this field have ushered in a formalism using the path integral method, whereby the behavior of a system of interest which is coupled to external systems may be studied in terms of its changed variables only. This is achieved by including the effects of the external system in a general class of functionals, called influence functionals of the coordinates, introduced by Feynman and Vernon [28]. Interesting properties of these influence functionals, which contain the entire effect of the environment including the change in behavior of the environment resulting from the reaction of the original system, have been studied in detail and have led to remarkable conclusions governing quantum systems interacting with linear dissipation systems (see Vasudevan [29]). Variational procedures have been developed to compute the functional integrals; path probability methods have been used in the study of irreversible statistical dynamics [30, 31], and interesting results have been achieved or re-established.

Questions regarding the utility of the path integral method have been raised, since all the results obtained up to now and using this method can be arrived at by other methods. Moreover, mathematical computation is also rather difficult. However, the viewpoint of the path probability methods restores to the physicist the conceptual advantages of classical mechanics in which we imagine particles following definite trajectories with certain probabilistic weights attached to these paths; while, in the more usual quantum mechanical apparatus this pictorial representation is replaced by an abstract mathematical formalism. For an excellent summary of these methods and their applications, the reader is referred to the review article by S. G. Brush [26].

REFERENCES

1. M. Born, *Z. fur Physik*, **37**(1926), 863.
2. M. Born, *Z. fur Naturforsch.*, **119**(1927), 354.
3. E. Wigner, *Phys. Rev.*, **40**(1932), 749.
4. R. P. Feynman, *Rev. Mod. Phys.*, **29**(1946), 367.
5. Alladi Ramakrishnan and R. Vasudevan, *Proc. Summer School in Theoret. Physics, Mussoorie, India*, Ministry of Scientific Research and Cultural Affairs, New Delhi, 1960.
6. Alladi Ramakrishnan and N. R. Ranganathan, *J. Math. Analysis Appl.*, **3**(1961), 261.
7. T. W. Marshall, *Proc. Roy. Soc. (London)*, **A276**(1963), 473.
8. R. P. Feynman and A. R. Hibbs, *Quantum Mechanics and Path Integrals*, McGraw-Hill, New York, 1965.
9. P. A. M. Dirac, *Principles of Quantum Mechanics*, Clarendon Press, Oxford, 1935. Sec. 33.
10. W. Heisenberg, *Physical Principles of Quantum Theory*, University of Chicago Press, Chicago, Ill., 1930, Chap. IV.
11. N. Wiener, *Act. Math.*, **55**(1930), 117.
12. M. Kac, *Trans. Amer. Math. Soc.*, **65**(1949), 1.
13. W. Montroll, *Comm. Pure Appl. Math.*, **5**(1952), 415.
14. I. M. Gelfand and A. M. Yaglom, *J. Math. Phys.*, **1**(1960), 48.
15. N. Saito and M. Namiki, *Progr. Theoret. Phys.*, **16**(1956), 71.
16. K. Ito, *Proc. Math. Statist. and Probability, Fourth Berkeley Symp.*, University of California Press, Berkeley, 1961, Vol. 2, p. 227.
17. R. H. Cameron, *J. Math. Phys.*, **39**(1926), 126.
18. J. G. Gilson, *Nuovo Cimento*, **40**(1965), 126.
19. J. G. Gilson, *J. Appl. Probability*, **5**(1968), 375.
20. H. Mori, I. Oppenheim and J. Ross *Studies in Statistical Mechanics*, North-Holland Publishing Co., Amsterdam, Vol. 1, 1962 part C.
21. J. E. Moyal, *Proc. Cambridge Philos. Soc.*, **45**(1949), 99.
22. H. Davison, *Proc. Roy. Soc. (London)*, **A247**(1958), 225.
23. W. K. Burton and A. H. de Borde, *Nuovo Cimento*, **2**(1955), 197.
24. Alladi Ramakrishnan, *Elementary Particles and Cosmic Rays*, Pergamon Press, London, 1962, Chap. II.
25. R. Abe, *Busserior Kenkyn*, **79**(1954), 101.
26. S. G. Brush, *Rev. Mod. Phys.*, **33**(1961), 72.
27. R. P. Feynman, *Phys. Rev.*, **91**(1953), 1291.
28. R. P. Feynman and F. L. Vernon, *Ann. Phys.*, **24**(1963), 118.
29. R. Vasudevan, *Symposia on Theoretical Physics*, Plenum Press, New York, Vol. 2, 1966.
30. R. Kikuchi, *Ann. Phys.*, **10**(1960), 127.
31. R. Kikuchi, *Phys. Rev.*, **124**(1961), 1691.

AUTHOR INDEX

Numbers in parentheses indicate the numbers of the references when these are cited in the text without the names of the authors.

Numbers set in *italics* designate the page numbers on which the complete literature citations are given.

161

SUBJECT INDEX

1996